A Quick Guide to Pipeline Engineering

QG Publishing is a Matthews Engineering Training Ltd company

MATTHEWS
ENGINEERING TRAINING LTD
www.matthews-training.co.uk

Training courses for industry

- Plant in-service inspection training
- Pressure systems/PSSR/PED/PRVs
- Notified Body training
- Pressure equipment code design ASME/BS/EN
- API inspector training (UK) : API 510/570/653
- On-line training courses available

Matthews Engineering Training Ltd provides training in pressure equipment and inspection-related subjects, and the implementation of published codes and standards.

More than 300 classroom and hands-on courses have been presented to major clients from the power, process, petrochemical and oil/gas industries.

We specialise in in-company courses, tailored to the needs of individual clients.

Contact us at enq@matthews-training.co.uk Tel: 07732 799351

Matthews Engineering Training Ltd is an Authorized Global Training provider to The American Society of Mechanical Engineers (ASME)

www.matthews-training.co.uk

A Quick Guide to

Pipeline Engineering

D. Alkazraji

BEng, CEng, MIMechE

Series editor: Clifford Matthews

Matthews Engineering Training Limited
www.matthews-training.co.uk

WOODHEAD PUBLISHING LIMITED

Cambridge England

Published by Woodhead Publishing Limited, Abington Hall, Abington
Cambridge CB21 6AH, England
www.woodheadpublishing.com
and
Matthews Engineering Training Limited
www.matthews-training.co.uk

First published 2008, Woodhead Publishing Limited and Matthews
Engineering Training Ltd
© 2008, D. Alkazraji

British Library Cataloguing in Publication Data
A catalogue record for this book is available from the British Library.

ISBN 978-1-84569-490-6   (book)
ISBN 978-1-84569-491-3   (e-book)

Typeset by Data Standards Ltd, Frome, Somerset, UK

Cover photographs by Anne and Keith Newton.

Contents

The Quick Guide Series

The *Quick Guide* data books are intended as simplified, easily accessed references to a range of technical subjects. The initial books in the series were published by The Institution of Mechanical Engineers (Professional Engineering Publishing Ltd), written by the series editor Cliff Matthews. The series is now being extended to cover an increasing range of technical subjects by Matthews Engineering Publishing.

The concept of the Matthews *Quick Guides* is to provide condensed technical information on complex technical subjects in a pocket book format. Coverage includes the various regulations, codes and standards relevant to the subject. These can be difficult to understand in their full form, so the *Quick Guides* try to pick out the key points and explain them in straightforward terms. This of course means that each guide can only cover the main points of its subject – it is not always possible to explain everything in great depth. For this reason, the *Quick Guides* should only be taken as that – a quick guide – rather than a detailed treatise on the subject.

Where subject matter has statutory significance, e.g. statutory regulation and referenced technical codes and standards, then these guides do not claim to be a full interpretation of the statutory requirements. In reality, even regulations themselves do not really have this full status – many points can only be interpreted in a court of law. The objective of the *Quick Guides* is therefore to provide information that will add to the clarity of the picture rather than produce new subject matter or interpretations that will confuse you even further.

If you have any comments on this book, or you have any suggestions for other books you would like to see in the *Quick Guide* series, contact us through our website: www.QGpublishing.com

Cliff Matthews
Series Editor

Invitation to New Authors

If you have an idea for a *Quick Guide* book and are interested
in authorship, we are interested in hearing from you. You do
not have to already be a published author (in fact we are
actively interested in finding new unpublished ones with a bit
of talent). All we ask is that:

- you know your subject as well as (or even slightly better
  than) others in your field;
- you can explain it in simple terms;
- you have the tenacity to write 25 000–30 000 words (and
  accompanying figures and tables).

If you can meet these requirements, then get in touch and we
will discuss with you the procedure for submitting your
proposal. Contact Cliff Matthews at:

author@QGpublishing.co.uk

Preface

Pipeline engineering is a large subject area covering a range of topics. This book provides a handy reference guide on both onshore and offshore pipeline engineering that engineers and students will find useful. Basic principles such as design, construction, operation and maintenance are discussed with the aim of being concise and informative. When working in the pipeline industry, there are numerous pipeline codes and standards, calculation approaches and reference material that the operator must understand in order to make accurate and informed decisions.

The book is divided into a number of sections including design, construction, risk assessment, pressure testing, operation and maintenance, condition monitoring, decommissioning and pipeline industry developments. Throughout this book, alongside these basic principles, there is reference to the main standards and literature that are used in the pipeline industry. These references are essential for further information. The book provides engineers and students with up-to-date and accurate information on current best practice and the underlying principles of pipeline engineering. For example, the engineer might need to know what the main corrosion assessment approaches today are, what quantitative risk assessment is or what methods are available for permanent and temporary repair. These are questions that I have put to myself and that have prompted me to produce a quick guide covering the full life cycle of pipelines, both onshore and offshore.

I would like to thank the following for their assistance in producing this book:

British Standards Institute (BSI)
American Petroleum Institute (API)
Institution of Gas Engineers and Managers (IGEM)

American Society of Mechanical Engineers (ASME)
Pipeline Research Committee (PRCI)
Det Norske Veritas (DNV)
Rosen Group
Advantica
Ameron BV
Corus Tubes Energy Business
Elsevier
NACE International

Finally, on a personal note I would like to thank Suzanne for her help and Cliff Matthews for checking my drafts and also for providing help and guidance on writing this 'Quick Guide'.

About the author

Working as a pipeline engineer, Duraid Alkazraji is a chartered engineer currently providing training services to Matthews Engineering Training. Duraid Alkazraji has worked in the pipeline industry for a number of years, having joined the BG group on their graduate development programme, and then going on to work for Advantica, PII and Saipem UK.

Summary

This quick guide to pipeline engineering covers a number of subjects:

- **The initial stage of pipeline design** is conducted before any work commences on the construction. It is important that environmental and legal considerations are taken into account, for instance an environmental impact assessment should be carried out to satisfy appropriate authorities and regulations. Detailed design can then be started. Firstly, the diameter and inlet pressure are decided upon according to the maximum acceptable pressure drop along the length of the pipeline. This is calculated using appropriate flow equations for gas or liquid flow. Other design parameters will then follow, including choosing an appropriate wall thickness, material grade and coating method. Finally, the maximum allowable operating stress will be decided upon according to the location of the pipeline route. Codes such as ASME B31.4 and B31.8 provide maximum operating stress limits based on the surrounding population density.

- The next stage then looks at **pipeline manufacture and construction**. Spools are manufactured using four main methods, with each manufacturing method generally varying in the pipe sizes available. These manufacturing methods include seamless, electric resistance welding (ERW), longitudinally submerged arc welding (LSAW) and spiral submerged arc welding (SSAW). Construction and land preparation stages must be started, which in the case of onshore pipelines involves a working corridor being created and a pipeline trench being dug to a depth of approximately 1.1 m. In the case of offshore pipelines, the trench is prepared using a dredger. Finally, positioning of the pipeline can be done using S-lay and J-lay methods for offshore pipelines. Once the pipeline is in the trench, it must be protected from corrosion using sacrificial or impressed current corrosion protection systems.

- **Controlling the risks from failure** is an important part of any integrity and management strategy. Consequently, regulatory bodies such as the Health and Safety Executive (HSE) within the UK and the Department of Transport (DOT) within the United States ensure that these management systems are in place. The two main risk assessment approaches used throughout the pipeline industry include quantitative and qualitative methods.

- Following these earlier stages, the pipeline is ready to begin **operation**. Before the pipeline can be operated safely, pipeline design codes require that it is pressure tested. A general hydrotest is usually conducted to 1.5 times the design pressure. The hydrotest will identify defects that may fail at the design pressure. Current best practice is to utilize high-level pressure testing, which provides a hydrotest safety margin. The basic principle of this is that the higher the pressure test used, the smaller the defects that will remain.

- There are numerous types of **defect** that may be found in pipelines, such as internal corrosion, external corrosion, laminations, stress corrosion cracking (SCC), cracks, dents and gouges. Consequently, it is important that pipeline operators utilize the most appropriate inspection methods available. Currently, the most widely used methods include magnetic and ultrasonic inspection tools, but specialist inspection tools are available that can detect SCC, channelling corrosion and seam weld defects.

- Ultimately, the engineer has to decide whether a pipeline containing a reported defect is fit for the intended pressure or whether it needs repair. There are various defect assessment approaches, all of which can be rather confusing, such as ASME B31.G, simplified RSTRENG and DNV-RP-F101, etc. The main methods used throughout the industry have been summarized in this book, including effective area methods, UTS-based corrosion assessment, dent fatigue life estimation and how to assess

planar defects such as cracks and laminations using codes
BS 7910 and API 579.

- Finally, **ongoing maintenance of pipelines** should then form
  part of a failure prevention and corrosion monitoring
  strategy including ongoing repairs and surveying techniques. Operators often have a number of tools available
  such as CIPPS, DCVG and Pearson Surveys. This book
  describes the capabilities of each method and the various
  permanent and temporary repair methods used throughout
  the industry.

Chapter 1

Principles of Pipeline Design

There are a number of important stages in the life cycle of an oil or gas transmission pipeline: design, construction, operation and maintenance and finally repair. This chapter will look at the initial stage of pipeline design for oil and gas pipelines. Within the planning phase, and before any work commences on constructing a new pipeline, factors that affect the design process include:

- the effect on the environment;
- the pipeline routing process;
- approval and legal considerations.

There are currently numerous standards available that provide guidance on the design of pipelines. Some operators may use their own national standard, but many others use foreign standards that are widely used throughout the pipeline industry. In particular, for oil and gas pipelines worldwide, the API (American Petroleum Institute), ANSI (American National Standards Institute) and BS (British Standards) are widely used. Within the UK, oil and gas pipelines are based on guidance provided by PD 8010 [1]. In addition, the IGE/TD/1 standard [2] is a pipeline code developed by the Institution of Gas Engineers and Managers within the UK for the design, construction and operation of pipelines operating at pressures exceeding 16 bar. In addition, IGE/TD/1 takes into account extensive research into the causes and consequences of pipeline failure. It is appropriate, therefore, that IGE/TD/1 be referenced for developments in international pipeline standards and current best practice throughout the pipeline industry. A summary of the main standards used worldwide includes those shown in Tables 1.1 and 1.2 [3].

Table 1.1   Overview of standards that provide guidance on design, construction and maintenance

Onshore

ASME B31.4	Oil pipelines
ASME B31.8	Gas pipelines
IGE/TD/1	Gas pipelines
PD 8010	Oil and gas pipelines
AS 2885	Oil and gas

Offshore

DNV Recommended Practice	Oil and gas pipelines
PD 8010	Oil and gas pipelines
API RP 1111	Oil pipelines
ASME B31.4	Oil pipelines
ASME B31.8	Gas pipelines

1.1 Effect on the environment

Consideration must be given to the likely impact a newly constructed pipeline will have on the environment. For onshore pipelines, these effects are highlighted in Fig. 1.1. For offshore pipelines, these effects are highlighted in Fig. 1.2.

It is important to identify the likely environmental effects of a planned pipeline and satisfy appropriate legislation. Obviously there will be different requirements around the world, but a typical example used within the UK includes the

Table 1.2   Overview of standards that provide guidance on design, construction and maintenance

Governing regulation	Design code	Approval organization
UK Pipeline Safety Regulations	PD 8010	HSE (Health and Safety Executive)
US Department of Transportation Regulations	DOT	DOT (Department of Transportation)

'Public Gas Transporter Pipeline Works Regulations'. This legislation requires an environmental impact assessment (EIA) in sensitive areas. Consequently, before the operator can construct a new pipeline, an EIA should be conducted at the design stage.

There are numerous environmental regulations and legislation available. Those that affect the UK pipeline industry include:

European Union Legislation
- 97/11/EC Effects of Projects on the Environment;
- 92/43/EEC Conservation of Natural Habitats of Wild Fauna and Flora.

UK Regulations
- The Gas Act and 'Public Gas Transporter Pipeline Works Regulations';
- Pipeline Works Environmental Impact Assessment Regulations 200, No. 1928.

UK Acts
- Environmental Act 1995;
- Environmental Protection Act 1990;

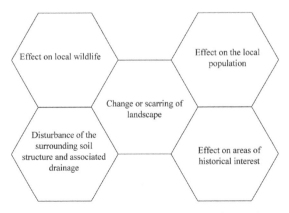

Figure 1.1   Environmental considerations for onshore pipelines

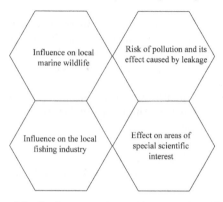

Figure 1.2 Environmental considerations for offshore pipelines

- Pipelines Act, for onshore oil pipelines.

For offshore pipelines, consideration should also be given to sensitive areas associated with new pipeline construction. This is controlled by:

- Petroleum and Submarine Pipelines Act [4], which governs the commercial extraction of oil and the protection of the environment.

1.2 Routing

Routing is an important factor in any pipeline design process as this determines areas through which the pipeline can and cannot be routed. If designing a pipeline from A to B, ideally it would be convenient to use the shortest route as a straight line between the two points (see Fig. 1.3). This is not always possible as, when routing onshore pipelines, the route must take into consideration:

- sensitive areas (national parks, forest regenerative areas);
- environment (wildlife, archaeological sites);
- geography (rocky areas, fault lines, areas of erosion);
- crossings (road, rail, rivers);

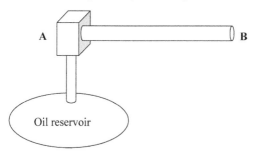

Figure 1.3   Routing of transmission pipelines

- areas of population;
- location of compressor stations;
- location of above-ground installations.

In addition, the route of the pipeline should avoid running parallel with high-density traffic routes, electricity power lines and other oil or gas pipelines. If a crossing is required, the pipeline should cross perpendicular to the road/railway (see Fig. 1.4). In addition, a feasibility study taking into account the above factors is required when defining the route between the start and end points of the pipeline. Usually, a 1 km wide strip is used to identify any deviations required in the pipeline route, as shown in Figs 1.4 and 1.5.

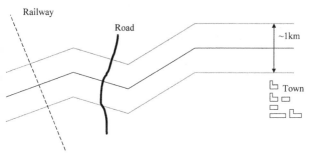

Figure 1.4   Pipeline routing corridor

Figure 1.5 Transmission pipeline routing

When routing offshore pipelines, again the straightest route is not always possible. Obstacles have to be avoided, such as:

- other offshore platforms;
- shipwrecks;
- subsea wellheads;
- underwater surface features (cliffs, volcanoes, erosion).

Care must be taken with routing pipelines that cross areas of high shipping activity, since there is the risk of damage to the pipe from anchors, or even the sinking of small boats if these become hooked on the pipe. The general principle is that the pipeline should be constructed so that it crosses perpendicular to the shipping lane, i.e. to ensure vessels travel the least amount of time within the region of the pipeline. Other considerations that may affect the pipeline route include those shown previously in Fig. 1.2. A seismic sonar survey is usually conducted to determine the overall terrain below the surface.

When looking at detailed design of offshore pipelines, there are a number of considerations, in particular:

- On bottom stability of the pipeline, hydrodynamic loads should be assessed to check whether the pipeline will be stable on the seabed under its own weight and will not move as a result of current and wave movement. These loads include lift, drag and inertia forces.
- Subsea pipelines that are not trenched are also susceptible to spans forming owing to surface erosion and movement. A consequence of this is that, as the current flows around the pipeline circumference, vortices are shed. As the frequency of these vortices approach the natural frequency of the pipeline, in-line vibration and vortex-induced vibration can occur. This results in fatigue of the pipeline and can ultimately cause failure. Analysis of free spans along the pipeline route should be conducted to check for vibration.
- Finally, subsea pipelines are also at risk from lateral and upheaval buckling. Usually this is complex and requires detailed modelling using finite element analysis software.

Table 1.3   Detailed design parameters

Pipeline diameter
Calculating wall thickness
Material grade
Maximum operating pressure and flow conditions
Operating temperature
Pressure drop
Corrosion protection

All these points must be taken into consideration when determining the route of a pipeline.

1.3 Approval and legal considerations

Having considered the environmental impact and routing selection, the next important stage is to notify the relevant authorities of the intention to construct a new pipeline. In the UK, authorization would be provided by the Department of Trade and Industry (DTI), who must be notified of any new construction projects and updated on the likely environmental effects. For cross-country pipelines, farmers should be consulted, since compensation payments are likely, in order to allow a pipeline to cross private land. In addition, permission will be required in areas where the proposed pipeline route will cross roads, railways or river crossings. Finally, to prevent any disruption to the project at the construction stage, appropriate measures should be taken to ensure that the proposed route does not affect protected wildlife species, preventing costly delays later in the project.

Once all these considerations have been addressed and the route options for the pipeline have been selected, detailed design of the pipeline system can be done. What does the detailed design involve? This includes looking at the detailed design parameters shown in Table 1.3, and is covered in detail in the following chapter.

Chapter 2

Design Approach

Having defined the pipeline route, taking into consideration factors described in Chapter 1, the next stage is to start the detailed design of the pipeline, including parameters such as volume throughout, length of the pipeline and acceptable pressure loss. Note that the length of transmission pipelines varies considerably and can range from less than 1 km to thousands of kilometres.

When deciding the form of product to transport, it is important to consider the advantages and disadvantages of using liquid or gas. The main advantages of liquid transmission pipelines include the following:

- During inspection using intelligent pigs, the speed is easier to control.
- The pipelines are easier to inspect using ultrasonics.
- It is possible to transport products in batches.
- Liquid is incompressible, and so the consequence of failure is less critical (i.e. flow can quickly be stopped).
- Flow is more controllable.

Disadvantages of liquid pipelines are as follows:

- There is a greater risk of pollution when leaks occur, i.e. hydrocarbons are heavier than air.
- Pipelines can easily become clogged with waxy deposits.
- There is a greater risk of corrosion from 'sour' operating conditions.

The main advantages of operating gas transmission pipelines include the following:

- Pollution is less critical since gases such as methane are lighter than air and diffuse into the atmosphere.
- Gases can easily be vented.

- Generally, gas pipelines suffer less from deposits than liquid pipelines.
- 'Sour' corrosion is not as big a problem as on liquid pipelines.

Disadvantages of gas pipelines are as follows:

- The consequence of failure is higher since the gas is compressible and flow is not as easily controlled.
- Inspection using ultrasonic tools is more complicated and specialist tools are required.
- Gas pipelines are usually operated as a single product.
- During inspection using intelligent pigs, the speed is more difficult to control owing to the compressible nature of gas.

2.1 Factors that influence the length of a pipeline

These factors are design pressure, acceptable pressure drop, diameter, wall thickness and temperature profile. The most

Liquid flow

Head loss $\dfrac{P}{\rho} = \dfrac{fLv^2}{2D}$

For laminar flow ($Re < 2100$) $\quad f = \dfrac{64}{Re}$

Reynolds number $\quad Re = \dfrac{\rho D v}{\mu}$

For turbulent flow ($Re > 4000$)

Colebrook–White equation $\sqrt{\dfrac{1}{f}} = -4\log\left[\dfrac{k}{3.7D} + \dfrac{1.413\sqrt{1/f}}{Re}\right]$

f = friction factor
L = length, (m)
k = surface roughness (m)
Re = Reynolds number
D = internal diameter (m)
v = fluid velocity (m/s)
$\bar{v}$ = average velocity (m/s)
μ = kinematic viscosity (cSt)
P = pressure (Pa)
ρ = density (kg/m³)

Gas flow

Gas flowrate equations (fully turbulent)

Weymouth

$$Q = 432.7 \frac{T_b}{P_b} \left[\frac{P_1^2 - P_2^2 - E}{GLT_{av}Z_{av}}\right]^{1/2} \cdot D^{2.667}$$

Panhandle

$$Q = 737.02 \left(\frac{T_b}{P_b}\right)^{1.02} \cdot \left[\frac{P_1^2 - P_2^2 - E}{G^{0.961}LT_{av}Z_{av}}\right]^{0.510} \cdot D^{2.53}$$

Q = flowrate, SCFD
P_b = base pressure (Pa)
T_b = base temperature (°C)
G = gas specific gravity
L = line length (m)
Z = compressibility factor
E = efficiency factor
D = pipe inside diameter (m)

Reprinted from McAllister, E.W. *Pipeline Rules of Thumb Handbook*, 6th edition, 2005, p. 314 (Gulf Professional Publishing), with permission from Elsevier.

Figure 2.1   Equations used in calculating the basic flow parameters for gas and liquid pipelines

commonly used equations for flow calculations are the Panhandle and Weymouth methods for gas pipelines. When considering liquid pipelines, the significant factors influencing pressure drop include the Reynolds number (Re) and the friction factor calculated using the Colebrook–White equation or from the Moody diagram. Figure 2.1 shows the basic equations used to calculate the pressure drop for gas and liquid pipelines [5]. It should be noted that, since gas is compressible, this requires more complicated analytical solutions for gas flow. Using these flow equations, the required diameter and operating pressure can be calculated.

2.2 Choosing a wall thickness for the pipeline

Having calculated the required pressure and diameter based on the above flow equations, it is important that the pipeline is thick enough to contain this design pressure. The basic formula shown in many pipeline codes, relating nominal wall thickness to design pressure for a straight section of steel pipe, is given in Fig. 2.2, where t is the nominal wall thickness, P is the design pressure (N/mm^2), D is the

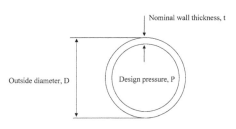

When designing pipelines a method of relating nominal wall thickness to design pressure for a straight section of steel pipe is:

$$P = \frac{2St}{D} \cdot FET$$

P = Internal pressure, N/mm^2
D = Diameter, mm
S = SMYS Specified Minimum Yield Strength, N/mm^2
F = Design factor
E = Joint factor
T = Temperature derating factor

Figure 2.2   Wall thickness calculation

diameter (mm), S is the specified minimum yield strength (SMYS), (N/mm^2), F is the design factor, E is the joint factor and T is the temperature derating factor.

Codes such as IGE/TD/1[2], PD 8010 [1, 6], B31.4 [7] and B31.8 [8] for transmission pipelines use this approach in calculating nominal wall thickness. When considering wall thickness for offshore pipelines, the pipe must be thick enough to prevent hydrostatic collapse under external pressure but also contain the internal pressure, and consequently the minimum wall thickness t_{min} is calculated as follows:

$$t_{min} = \frac{(P_{int} - P_{ext})D}{2FSET} + CA$$

where CA is the corrosion allowance and P_{int} and P_{ext} are internal and external pressure respectively.

2.3 Choosing an appropriate material grade for the pipeline

The material properties are important in any pipeline system since it is important that the pipeline does not yield under stress or fail owing to fracture initiation and can also be easily welded. The basic relationship between these factors is shown in Fig. 2.3.

When deciding on the material to use for onshore or offshore pipelines, it is important to ensure that the steel has adequate strength, fracture toughness and weldability. Consequently, steel is often ordered to specification using codes such as the American Petroleum Institute Standard API 5L [9] or the European Standard EN10208-2 [2]. Tables 2.1 and 2.2 show the range of materials available [2]. It should be noted that these are minimum yield and tensile strength values according to specification, and that the actual measured strength values are normally higher.

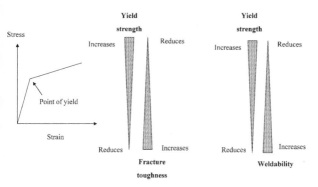

Figure 2.3   Relationship between material properties and weldability

2.4 Toughness

Toughness is defined as the resistance of a material to crack propagation. If a pipeline were made from low-toughness material, this would be at risk of brittle facture and would not be able to tolerate cracks. Indeed, the tougher the material, the larger the crack that can be withstood. The

Table 2.1   Material specifications according to API 5L

	API 5L	
Grade	SMYS (N/mm^2)	UTS (N/mm^2)
Grade A	207	331
Grade B	241	413
X42	289	413
X46	317	434
X52	358	455
X56	386	489
X60	413	517
X65	448	530
X70	482	565
X80	551	620

Table 2.2   Material specifications according to EN 10208-2

| Grade | EN 10208-2 | |
	SMYS (N/mm²)	SMTS (N/mm²)
L245	245	415
L290	290	415
L360	360	460
L415	415	520
L450	450	535
L485	485	570
L555	555	625

main factors that affect toughness level include operational temperature, geometry and operational stress. Consequently, operating at higher stress levels requires higher toughness material, whereas operating at lower stress levels permits lower toughness levels. Material specifications such as API 5L [9] provide recommendations on minimum toughness requirements. In addition, codes such as B31.8 [8] and IGE/TD/1 [2] recommend fracture control methods by considering fracture appearance on full-sized Charpy specimens. This includes:

- *Brittle fracture control.* To protect against brittle fracture propagation, the Battelle drop weight tear test (BDWTT) can be used.
- *Ductile fracture arrest.* To protect against ductile fracture propagation, the Charpy V-notch impact test can be used.

2.5 Operational pressure

Pipeline failure is a risk that the designer must take into account. A method of minimizing this risk is to set a maximum operating stress limit based on the level of population within the vicinity of the pipeline. This is known as 'area classification', and many codes set a design factor, which prevents the pipeline from operating above certain stress levels. These design factors apply to pipelines ranging from risers to onshore pipelines. Typical classifications and stress levels are summarized in Tables 2.3 and 2.4 and are based on PD 8010 [6], ASME B31.4 [7], ASME B31.8 [8] and CSA Z662 [10].

Design factors (DFs) provide a safety margin to ensure

Table 2.3 Location classifications and corresponding design factors for onshore pipelines

	PD 8010		ASME B31.4		ASME B31.8		CSA Z662	
Class	Design factor	Definition (people/ha*)	Design factor	Definition (people/ha*)	Design factor	Definition (buildings)	Design factor	Definition (buildings)
Class 1	0.72	<2.5 rural	0.72	None	0.8 (Div. 1) 0.72 (Div. 2)	≤ 10	0.72	≤ 10
Class 2			0.72	None	0.6	>10 and <46	0.6	>10 and <46
Class 3	0.3	>2.5 semi-rural	0.72	None	0.5	≥46	0.5	≥46
Class 4	0.3	Central town	0.72	None	0.4	Multistorey buildings > four stories	0.4	Multistorey buildings > four stories

*ha = hectare

Table 2.4 Design factors for offshore pipelines

	Risers and platform piping	Seabed pipeline
ASME B31.4		
Hoop stress	0.60	0.72
Longitudinal stress	0.80	0.80
Combined stress	0.90	0.90
ASME B31.8		
Hoop stress	0.50	0.72
Longitudinal stress	0.80	0.80
Combined stress	0.90	0.90

that a pipeline does not operate at 100% SMYS, and are based on the following equation:

$$DF = \frac{\sigma_H}{SMYS}$$

where σ_H is the hoop stress and SMYS is the specified minimum yield strength.

The hoop stress uses a formula known as the Barlow equation and is quoted in many of the pipeline codes (see Fig. 2.4). It must be noted that for offshore pipelines the external hydrostatic pressure should also be taken into account, which acts against the internal pressure on the pipe wall.

2.6 Temperature effects

If the pipeline is to operate under extreme high or low temperatures, material properties will change and affect operational conditions, so they need to be taken into account. Most pipelines are used within the metal temperature limits according to specifications such as API 5L, but, when operating beyond these limits (high or low temperatures), consideration must be given to the resulting strength and toughness properties. The B31.8 code provides temperature-derating factors at different operational temperatures above 250 °F, since yield strength will be affected. Under extreme low temperatures, toughness properties of the steel

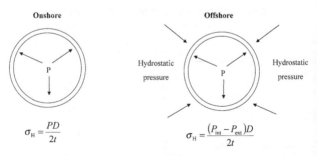

Figure 2.4   Calculation of hoop stress

change, so operating below the material brittle–ductile transition temperature would increase the risk of failure due to brittle fracture. IGE/TD/1 [2] recommends that material should be tested for adequate toughness properties at 0 °F.

2.7 Surge

In addition to the loads described, consideration must be given to surge pressure. Surge is caused by a sudden change in flow resulting in a pressure wave that travels through the fluid. Typical causes of surge would be:

- sudden valve closure;
- pump start-up;
- blockages in the pipeline.

2.7.1 Guidance provided by codes

Most design codes specify an allowable pressure margin for this overpressure. PD 8010 [6] and B31.4 [7] allow a maximum of 10% overpressure from the design pressure (see Fig. 2.5).

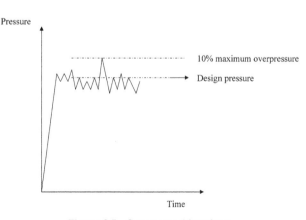

Figure 2.5 Surge considerations

2.8 Pipeline coating

Figure 2.6 shows pipe spools coated with fusion-bonded epoxy. In order to prevent corrosion occurring, the two main methods of pipeline protection are.

- pipeline coating;
- a cathodic protection (CP) system.

The primary method of protection against corrosion should always be pipeline coating, which must also be backed up by an effective cathodic protection system. Corrosion usually occurs at the site of coating defects where moisture gains access to the pipe surface. If the CP system is not effective enough, then corrosion will occur. Other sources of corrosion

Courtesy of BSR Pipeline Services Limited

Figure 2.6   Pipeline coating methods

include where the coating has started to form 'holidays', trapping moisture between the pipe surface and coating.

The pipeline coating must have a number of characteristics, such as:

- thermal stability (will not deform under high operating temperatures);
- impermeability to water and moisture;
- chemical stability (does not degrade owing to chemical reaction with soil or surroundings);
- ease of use;
- economical.

Figure 2.7 shows the typical coating methods for onshore and offshore pipelines. The most widely used coatings are summarised as follows.

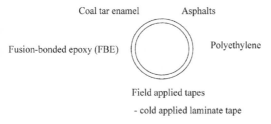

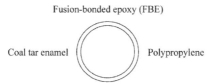

Figure 2.7 Typical pipeline coating methods

2.8.1 Coal tar enamel coatings

Coal tar enamel and bitumen coatings were often used on many of the older pipelines around the world. Application is made by wrapping a glass fibre tissue around the pipe circumference and saturating the coating with a melted mix of bitumen or coal tar. Disadvantages of this coating are as follows:

- It is sensitive to high or low temperature changes.
- It is susceptible to cracking due to soil stressing.
- Disbonding of the coating occurs through impact damage or from poor surface preparation on the pipe.

2.8.2 Tape coatings

Tape coatings are often used to repair pipe sections that have been excavated to repair areas of existing damaged coating. The most common types are hot applied tapes where a fabric coated with bitumen is applied around the circumference and heated. Alternatively, cold applied tapes include those made from polyethylene, which have a self-adhesive layer. While the tape coating method is relatively cheap, a disadvantage is that tape coatings are not tolerant to high operational temperatures and can be susceptible to soil stressing.

2.8.3 Heat-shrinkable plastic coatings

These are essentially plastic sleeves or sheets that are sensitive to heat. When they are placed around the pipe circumference and heat is applied from a blowtorch, this causes the plastic to contract and shrink onto the pipe surface. The method is generally used for smaller-diameter pipes, particularly around pipe joints/field joints.

2.8.4 Polyethylene coatings

Polyethylene coatings are applied during the manufacturing stage, either as an adhesive tape wrapped around the pipe circumference or in the form of a single extruded polyethylene coating. During the extruded polyethylene manufacturing process, an epoxy primer is applied, followed by the adhesive and finally one or two layers of polyethylene. The

Table 2.5   Relative cost of different coating methods

Coating method	Relative cost
FBE, polyethylene	High
Bitumen, coal tar, asphalt	Medium
Tape coatings	Low

temperature tolerance is not as high as that of fusion-bonded epoxy coatings since it is limited by the adhesive material.

2.8.5 Fusion-bonded epoxy (FBE) coatings

FBE is currently one of the most reliable coating methods and is often used by pipeline operators. Generally, this is applied during the manufacturing stage, but it can also be applied on-site on areas of damaged coating such as field joints. Application of the FBE coating is firstly through surface preparation and heating of the pipe surface prior to application of the epoxy powder. These fine particles melt over the pipe surface and form a strong bond once it is quenched. This form of coating is reliable and is resistant to coating defects, disbonding and higher operational temperatures.

Table 2.5 shows the relative cost each of the coating methods.

2.9 Pipeline protection

Methods of pipeline protection include:

- concrete coating;
- increased wall thickness;
- burial;
- sleeve protection;
- marker tapes;
- protective concrete slabs.

Areas typically requiring protection for onshore pipelines include road crossings, rail crossings or other sensitive areas. In the case of offshore pipelines these include areas of

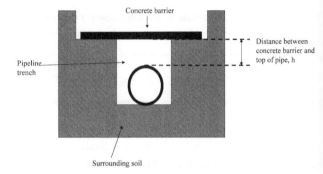

Figure 2.8 Pipeline impact protection

increased shipping activity. Concrete mattress protection is often used for offshore pipelines. This has a number of purposes including:

- weight coating (negative buoyancy);
- impact protection.

Since the pipeline is submerged, concrete coating provides negative buoyancy to prevent the pipeline from floating to the surface. In addition, this coating acts as impact protection during its operational life to protect against damage from:

- ship anchors;
- fishing equipment (trawling equipment).

For onshore pipelines, protection is usually done through markers or concrete slabs above the pipe. IGE/TD/1[2] recommends the use of concrete slabs to prevent impact damage, as shown in Fig. 2.8. In addition, the spacing, h, between the pipe and concrete must be large enough to prevent impact from a pneumatic jackhammer.

Chapter 3

Pipeline Construction and Risk Assessment Techniques

3.1 Pipeline manufacturing methods

The four main methods for pipe manufacture are:

- seamless pipe (pipes with no seam weld);
- electric resistance welded pipe (ERW);
- longitudinal-seam submerged-arc welded pipe (LSAW);
- spiral submerged arc-welded pipe (SSAW).

Seamless pipe usually comes in smaller sizes, 450 mm and below in diameter. The main methods of seamless pipe manufacture include rotary forging, plug mill and extrusion. One such manufacturer is the Mannesmann Mill, which produces seamless pipe in what is known as a 'plug mill'. This method uses a solid billet of steel that is formed into the shape of a pipe using a series of rollers and a piercing arm that forms the material into the shape of a tube. The tubes are then shaped into pipes where the wall thickness and pipe diameter measurements are controlled using a series of rollers and an internal forging arm (see Fig. 3.1).

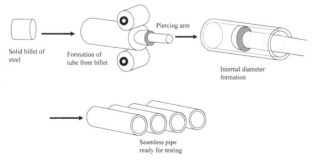

Figure 3.1   Seamless pipe manufacture

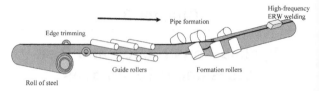

Figure 3.2   ERW pipe manufacture

Welded pipe is generally made by bending steel sheets into the shape of a pipe and welding them together. In the case of ERW pipe, the process starts with a rolled sheet of steel being uncoiled and straightened, and the edges being prepared ready for bending into the shape of a pipe. The bending is done by drawing a continuous flat sheet of steel through a series of pressure rollers and heating coils, forming the tube shape. Finally, the process of electric resistance welding is carried out by passing a high-frequency electric current through sliding contacts on the pipe surface, fusing the edges together to create a neat defect-free weld. This type of pipe is generally available in smaller diameter sizes of 500 mm and below (see Fig. 3.2).

Another type of welded pipe is the longitudinal sub-merged-arc welded pipe (LSAW). This construction method is summarized in Fig. 3.3. Generally, this is used for larger-diameter pipelines of 400 mm and above:

- Plates of steel are cut to the required pipe length (typically 12 m), and the longitudinal edges are prepared so that welding can be done.
- The edges of the plates are firstly crimped, then forced into a U-shape and finally forced into the circular O-shape using a series of cold pressing operations.
- Finally, the longitudinal seam is then welded internally and externally using submerged-arc welding (see Fig. 3.4).

To prove the integrity of the fabrication, a series of tests are carried out, including;

- A mechanical expansion test to create ~1.5% strain. This increases the strength and tests the integrity of the weld.
- A hydrostatic test to ~95% of the specified minimum yield strength (SMYS).
- Inspection using NDT such as ultrasonics, magnetic particle inspection or radiography.

Spiral submerged-arc welded (SSAW) pipes are used for larger-diameter pipelines of 400 mm and above, but the pipe diameter is dependent on the angle of coil and thickness of the sheet used. Manufacture is by using a hot roll of steel that is unwound using a series of rollers. During this unwinding process, the edge of the steel sheet is prepared for welding and the sheet is forced by the rollers into a coil shape. Finally, the edge is welded onto the trailing edge of the previous coil using submerged-arc welding. Like other manufacturing methods, the completed pipe sections are

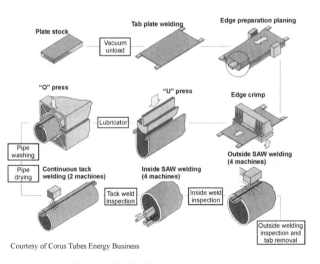

Courtesy of Corus Tubes Energy Business

Figure 3.3   SAW pipe manufacture

Courtesy of Corus Tubes Energy Business

Figure 3.4   Submerged-arc welding

hydrotested and finally tested using NDT in the mill to check the integrity of the weld.

3.2 Land preparation, excavation and pipe stringing

Land preparation of onshore pipelines involves a number of stages, as shown in Fig. 3.5. In summary, the main stages for onshore pipeline construction are as follows:

- obtaining the landowner's consent;
- preparation of a working corridor;
- stripping of topsoil;
- excavation of the trench;

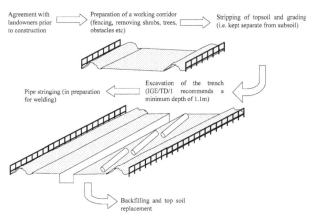

Agreement with landowners prior to construction → Preparation of a working corridor (fencing, removing shrubs, trees, obstacles etc) → Stripping of topsoil and grading (i.e. kept separate from subsoil)

Pipe stringing (in preparation for welding) ← Excavation of the trench (IGE/TD/1 recommends a minimum depth of 1.1m)

Backfilling and top soil replacement

Figure 3.5   Stages during land preparation

- pipe stringing;
- backfilling and topsoil replacement.

The main benefit of burying a pipeline is often to protect it from damage, but it also allows the landscape to be restored to its original form. Offshore pipelines can also be buried in a trench, or they can simply be laid on the sea floor.

The main stages for offshore pipeline construction and land preparation are as follows:

- survey of the terrain using side-scan sonar to determine the surface soil types and physical characteristics such as outcrops, boulders and holes;
- preparation of the route prior to installation, which involves presweeping of sand waves and removal of debris using a dredger;
- crossing of other pipeline routes, prepared by using specially designed mattresses or bridges.

Laying of the pipeline can be conducted using different methods. The main approaches are:

- *The 'S-Lay' process*, shown in Fig. 3.6. Single 12 m pipe sections are welded together and tested using NDT, and field-joint coatings are applied in sequence on a barge or vessel. This is then positioned on the seabed using a stinger attached to the end of the vessel and a tensioning system. This creates the S-shape as the pipeline is positioned on the seabed.

- *The 'J-lay' method* (see Fig. 3.7). This method uses a single workstation to produce multiple preassembled sections of pipe, which is then raised and positioned on the seabed using the J-shape lay curve. Benefits of this approach rather than using the S-lay method is that the pipeline is subjected to lower stress levels.

- *Bottom towing*. Bundles of the pipeline are dragged along the seabed to the required location.

- *Reel lay*. The pipeline is preconstructed onshore and then wound onto a large reel, which is then J-laid into position on the seabed.

Excavation of the trench to allow burial of the pipe below the sea floor is conducted using three main methods:

- *Cutting tool*. A machine drives along the pipeline route, cutting into the soil using mechanical teeth.

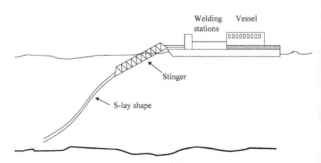

Figure 3.6 'S-lay' method

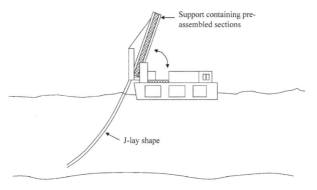

Figure 3.7 'J-lay' method

- *Ploughing*. A subsea ploughing machine is towed over the surface using a barge.
- *Jets*. These produce a high-pressure jet of water that blasts the soil away, creating the trench.

3.3 Corrosion protection

As described in Section 2.7, the main methods of pipeline corrosion protection are a pipeline coating and an appropriate cathodic protection (CP) system. The two main CP methods are:

- sacrificial protection system;
- impressed current protection system.

The basic elements of pipeline corrosion are shown in Fig. 3.8. For corrosion to occur, two areas must exist: the anode and the cathode. When a potential difference occurs between the anode and the cathode, and they are connected through the pipe surface with an electrolyte (i.e. surrounding soil), conditions exist to form a corrosion cell. Once this corrosion cell occurs, current flows away from the anode through the electrolyte to the cathodic area where current flows into the metal. Corrosion occurs only at the site of the anode where iron forms metal ions that react with the surrounding moisture, forming rust (see Fig. 3.9).

The corrosion rate in a corrosion cell is dependent on the potential difference between the anode and cathode. Relating this to pipelines, factors that affect the likelihood of a corrosion cell occurring are:

- Variations in soil type (i.e. electrolyte).
- Anaerobic or aerobic soil conditions (anaerobic soils tend to form the anodic area, whereas aerobic areas form the cathode).
- Dissimilar metals between pipe spools in a continuous pipeline system (reference should be made to the natural potential of different metals using the galvanic series). If two dissimilar metals are directly connected in an electrolyte, then the metal with the lower negative natural

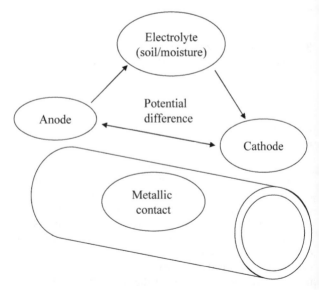

Figure 3.8 Basic elements of pipeline corrosion

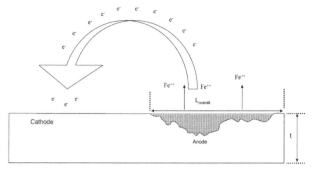

Figure 3.9   Anode and cathode areas

potential will corrode preferentially. For example, if an old brass pipe is connected to a new steel pipe, then the steel pipe will corrode in preference since this has a more negative natural potential.

3.3.1 Sacrificial current protection system

The basic principle of a sacrificial protection system is based on the natural potential of metals. Protection is gained by introducing another metal in the soil/electrolyte that will corrode in preference to the pipeline metal (i.e. is more electronegative). This metal then acts as the anode and is termed the sacrificial anode. Commonly used anodes include magnesium blocks that are placed alongside the pipeline and corrode in preference to the pipe surface. Applications include:

- subsea pipelines;
- short onshore pipeline sections.

3.3.2 Impressed current protection system

The impressed current protection system is generally used for long cross-country pipelines and uses mains electricity and a transformer to apply a d.c. voltage between the anode and cathode. External voltage is used to drive the protection current and ensures that the pipe surface is always the cathode. The two main measurements of this system are the 'on' and 'off' potentials, taken with reference to a copper sulphate electrode. Where possible it is important to measure the potential at the surface of the pipe (i.e. polarized potential), and consequently 'off' potential measurements are taken. This process is known as synchronous interrupting. When this is not possible, which may be the case in older protection systems, potential measurements should be taken at the surface using 'on' potential measurements.

3.3.3 Codified guidance on cathodic protection

Codified guidance such as NACE recommended practice [11] states that, for an effective CP system, the target 'off' potential level along the pipeline should be maintained within a potential range of $-850\,mV$ to $1200\,mV$. Where 'off' potentials cannot be measured, the 'on' potential should be maintained above $-1250\,mV$.

3.4 Pipeline codes and standards

The use of codes can be confusing since there are numerous available, and different countries have different national standards and codes for best practice. However, the main codes used throughout the pipeline industry for both oil and gas pipelines are shown in Fig. 3.10.

The basis of these codes is to provide guidance on the design, construction and operation of pipelines. Essentially, these ensure that the pipelines are designed and constructed safely and that the integrity of the pipeline is maintained throughout its life cycle, comprising the following stages:

- design;
- construction;
- pressure testing;
- operation;
- maintenance;
- repair;
- decommissioning.

Following the Piper Alpha disaster, the Health and Safety Executive (HSE) brought out the Pipeline Safety Regulations

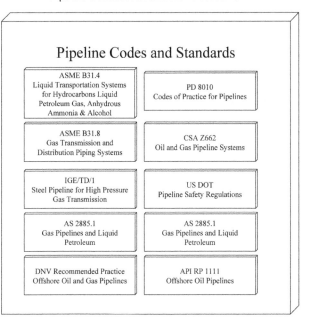

Figure 3.10 Worldwide pipeline codes and standards

(PSRs) [12]. The basic concept of the PSRs is summarized in Fig. 3.11.

The result of the Piper Alpha disaster was that major changes were made relating to safety in the oil and gas industry. This had an effect on both onshore and offshore pipelines. Under these new regulations, all high-pressure pipelines are considered as major accident hazard pipelines, placing responsibility on pipeline operators to ensure:

- adequate maintenance of pipeline integrity;
- a major accident prevention document (MAPD) is in place.

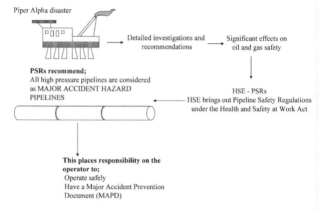

Piper Alpha disaster

Detailed investigations and recommendations → Significant effects on oil and gas safety

PSRs recommend;
All high pressure pipelines are considered as MAJOR ACCIDENT HAZARD PIPELINES

HSE - PSRs
HSE brings out Pipeline Safety Regulations under the Health and Safety at Work Act

This places responsibility on the operator to;
Operate safely
Have a Major Accident Prevention Document (MAPD)

Figure 3.11   Pipeline safety regulations

This MAPD is a guidance document demonstrating that the operator has considered all potential risks to the pipeline and has implemented a safety management system to control these risks. Changes were also made to the offshore pipeline industry, with the Submarine Pipelines Act (1975) being updated in 1998, preventing commercial exploitation at the expense of safety.

3.5 Risk assessment techniques

Pipeline operators have responsibility for maintaining a safety management system to control the risks of pipeline failure. Statutory bodies such as the DOT in the United States and the HSE within the UK have the responsibility to ensure that these management systems are implemented. These include:

- Pipeline Safety Regulations (PSRs) within the UK;
- Office of Pipeline Safety (OPS) within the United States.

As part of this overall safety management programme, risk assessments are conducted, which involve considering both the consequence of failure and the probability of failure (see Fig. 3.12).

The two main approaches include quantitative and qualitative risk assessment.

3.6 Quantitative risk assessment

With this approach, risk levels are quantified using direct calculation. UK legislation requires that risks are reduced to a level that is 'as low as reasonably practicable', or ALARP. Figure 3.13 shows the ALARP diagram used by the HSE to describe criteria for acceptable and unacceptable levels of risk [2].

The ALARP concept shows an unacceptable region where the risk of death cannot be justified, a broadly acceptable region where the risk levels are tolerable and a region where risk levels are insignificant. Quantitative risk assessment can therefore be used to justify exceptions to specific code recommendations such as described in IGE/TD/1[13,2]. By analysing the probability and consequences of failure, safety measures can be put in place to ensure that risks are set at an acceptable level.

Quantitative risk assessment involves conducting detailed calculations on individual risk and group or societal risk. Individual risk is defined as the frequency at which individuals may sustain injury from a particular hazard (see Fig. 3.14).

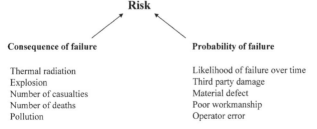

Risk

Consequence of failure	Probability of failure
Thermal radiation	Likelihood of failure over time
Explosion	Third party damage
Number of casualties	Material defect
Number of deaths	Poor workmanship
Pollution	Operator error

Figure 3.12   Elements of risk assessment

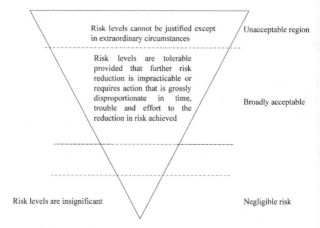

Figure 3.13 ALARP, 'as low as reasonably practicable'

Using the results of these calculations, criteria such as the ALARP diagram (UK) can then be employed to determine if the risk is acceptable. As shown previously in codes such as B31.8 [8], methods of controlling the consequence of failure include limiting the maximum operating stress level in certain locations (see Fig. 3.15).

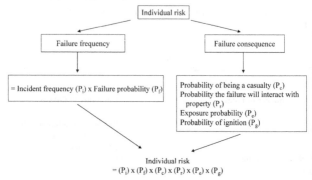

Figure 3.14 Quantitative risk assessment

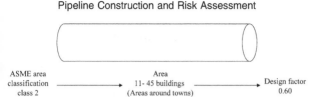

Figure 3.15 Controlling the consequence of failure

A common method used to represent the societal risk is through a plot of event frequency F causing a number of deaths N, shown in Fig. 3.16 [2]. Here, the graph is divided into a number of envelopes that define acceptable and unacceptable levels of societal risk.

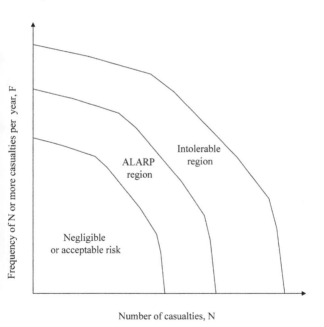

Figure 3.16 Failure frequency versus number of deaths

3.7 Qualitative risk assessment

A simpler approach of risk assessment is by means of qualitative risk assessment. Differences with this method are as follows:

- Unlike the quantitative approach, this is not based on numerical calculations, and uses a ranking system for consequence and probability of failure.
- The analysis approach is by means of a 'risk matrix'.

Using this matrix, risk is defined as probability of failure × consequence of failure, with a relative ranking system used to compare different pipeline sections (see Fig. 3.17).

The tool can be used as a ranking system for different pipelines by identifying the most likely failure mechanisms. This approach is the basis for risk-based inspection (RBI)

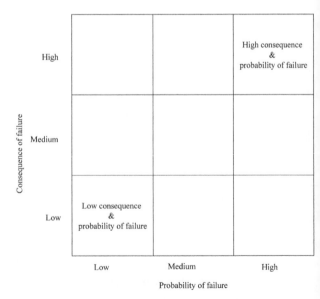

Figure 3.17 Qualitative risk assessment

and is considered an effective method of optimizing integrity costs among pipeline operators worldwide.

The main stages in developing an RBI programme are:

- identify failure mechanisms;
- conduct risk assessment on the pipeline(s);
- identify the required inspection method;
- define an inspection plan based on damage growth mechanisms and severity.

To illustrate this method, look at the following example. Figure 3.18 shows the first stage of identifying the likely failure mechanisms for three pipeline sections A, B and C and ranking them on the basis of consequence and probability of failure. In the case of pipeline A, this is subjected to high fatigue cycles, and previous inspections have identified cracks at seam welds. Failure of this type of defect would be by brittle fracture or plastic collapse, depending on the material properties. Consequently, the probability of failure for this type of defect is high. In addition, owing to the likely failure mechanisms, the consequence of failure is also high.

The next question to ask is whether this risk can be easily controlled through regular inspections. In this case, the answer is YES. Inspection tools are available that can detect crack-like defects at seam welds, so the value of regular inspection on this pipeline is high. This is highlighted in Fig. 3.19 where a third dimension on the matrix is added for inspection requirement. In this case, regular inspection would be instrumental in:

- controlling risk;
- preventing failure.

The next approach is to identify the available inspection methods:

- ultrasonic wall measurement tool;
- ultrasonic crack detection tool;
- transverse field inspection tool.

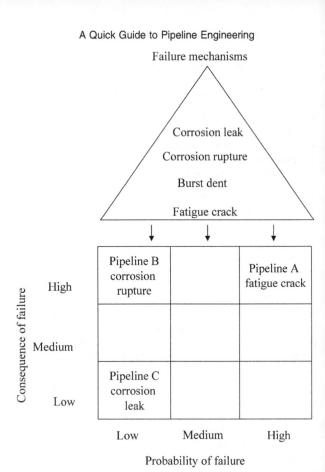

Figure 3.18   Example of qualitative risk assessment

Having identified the available tools and their sizing capabilities, the final step is to define an inspection plan. The inspection interval for different failure mechanisms will depend on:

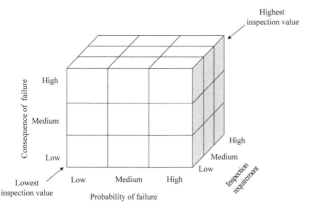

Figure 3.19   Risk-based inspection

- the growth rate of the defect;
- the choice of inspection method;
- the change in operating and loading conditions;
- the change in failure probability over time.

Figure 3.5   The ...

- ...
- ...
- ...
- ...

Chapter 4

Pressure Testing and Commissioning

4.1 Pressure testing

Many pipeline codes require that a pipeline is pressure tested. There are two main methods that pipeline operators use to test the integrity of a pipeline:

- A standard hydrotest to 1.5 times the design pressure – this provides an immediate test of integrity.
- A high-level pressure test – this provides a safety margin against growth of defects during operation.

Current best practice is to use liquid (water from a river, lake or other source) to test the pipeline because:

- It is more environmentally friendly.
- Water is incompressible, so if failure occurs the pressure is immediately dissipated and there is little risk of long propagating fractures. Conversely, gas is compressible and contains considerably more energy when pressurized, so pneumatic testing carries the risk of causing long propagating fractures, owing to the larger amount of stored energy within the pipe wall.

For pipelines operating in extremely dry environments such as the desert, pneumatic testing or intelligent inspection tools are used as an alternative to water pressure testing.

High-level pressure testing is an effective method of proving the integrity of a pipeline and removing critical defects at the time of commissioning. A lower-pressure hydrotest does not account for the possibility of defects growing under the influence of operating pressure. To provide a safety margin against failure of smaller defects, the basic principle of a high-level pressure test was developed. Figure 4.1 shows the benefits of conducting a hydrotest.

It is important to consider whether a defect was at the

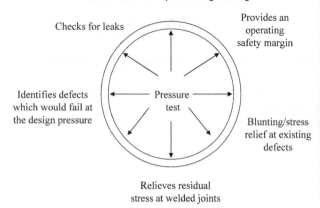

Figure 4.1 Why hydrotest pipelines?

point of failure when a pressure test was conducted. Two significant factors influence this:

- Existing defects that are in a state of tension at the crack tip end up in a state of compression due to plastic yielding. This restrains the crack tip and prevents the defect from growing.
- For a given size of defect and test pressure, there is a time period beyond which no further failures occur (i.e. those defects that would become critical during the operating life of the pipeline would fail during the test). A hold period of 24 h is recommended in codes such as IGE/TD/1 [2]. The main reason for this is to ensure that there are no leaks, and that no defects have failure mechanisms that are time dependent. Codes such as B31.8 [8] allow a shorter test duration of at least 2 h.

The basic principle of high-level pressure testing is:

- Smaller defects will remain as the test pressure is increased. This is the main principle of the hydrotest safety margin that many codes specify (see Fig. 4.2).

As shown in Fig. 4.2, the higher the level of the pressure test, the greater is the crack growth margin. Guidance and recommendations on pressure test levels are provided in different pipeline design codes; a summary of these recommendations is shown in Table 4.1. IGE/TD/1 [2] provides an equation for calculating the test pressure:

$$P_t = 20 \cdot t_n \cdot S \cdot f / D$$

where P_t is the test pressure (bar), t_n is the nominal wall thickness (mm), S is the SMYS, f is the fraction of SMYS for the test pressure and D is the outside diameter (mm).

As shown in Table 4.1, reference is made to the 'half-slope'. In test conditions, the pressure within the pipeline is monitored by plotting a graph of pressure versus volume. Figure 4.3 shows a plot of pressure against volume input.

Since pressure is increased using a pump, during pressurization, the number of pump strokes is counted to give a fixed

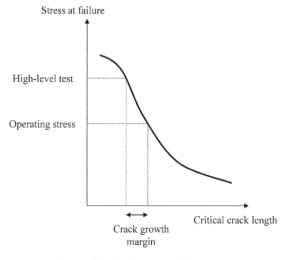

Figure 4.2   Hydrotest safety margin

Table 4.1   Recommended hydrotest levels from design codes

Design code	Hydrotest level	
PD 8010	150% of maximum allowable operating pressure (MAOP) or 90% SMYS	
B31.4	For operation where hoop stress is >20% SMYS, test must be 1.25 x MAOP	
B31.8	Class 1	
	(Div. 1)	1.25 x DF*
	(Div. 2)	1.1 x DF
	Class 2	1.25 x DF
	Class 3	1.4 x DF
	Class 4	1.4 x DF
	Offshore	1.25
IGE/TD/1 Gas	Operating above: 30% SMYS, lower of: (1) 105% SMYS or (2) Half-slope	Operating below: 30% SMYS (1) 1.5 x design pressure

* DF = design factor (operating stress/SMYS).

pressure increase ($\sim$7 bar/min). In addition, the volume of water used to produce this increase is also measured. As air is compressible, any trapped air will show during pressurization on this plot. Consequently, the air content is determined by projecting a straight line through the horizontal axis. When pressurizing, a 'half-slope' occurs when the number of pump strokes required to give a pressure increase doubles. This is known as 'double stroking'.

Factors that effect the hydrotest include:

- material properties of the pipeline;
- wall thickness;
- elevation changes;
- water availability and disposal.

The main benefit of hydrotesting a pipeline include leaving a compressive residual stress at the crack tip of a defect. During pressurization, the stresses in the pipe cause plastic deformation at the crack tip, creating a tensile stress in this region (see Fig. 4.4). On completion of the hydrotest, as pressure is reduced, the residual stress now acting at the

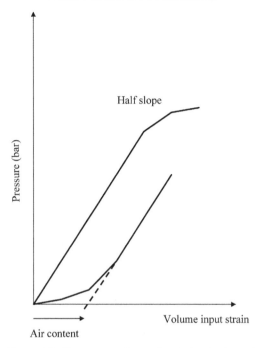

Figure 4.3   Pressure–volume plot during hydrotest

crack tip is compressive. This effectively acts as a restraint, preventing the crack from growing further, and can be beneficial to the pipeline.

Offshore pipelines have a number of components that need to be pressure tested, including:

* risers;
* valves;
* pig traps;
* subsea pipeline.

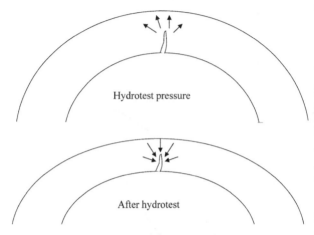

Figure 4.4   Stress state at a crack tip during and following the hydrotest

Since risers are in close vicinity to a platform, they are designed with thicker walls and have a lower design factor (Chapter 2, Table 2.4). In addition, the manufacturer conducts a separate pressure test to a higher level prior to installation. When all these sections have been constructed, the pipeline is flooded with seawater, treated with corrosion inhibitor and pressure tested.

The next stage following the hydrotest is 'commissioning', which involves preparing the pipeline for operation by drying or swabbing to remove any remaining water.

4.2 Commissioning

After the hydrotest, the next stage is to dry the pipeline to ensure that no water remains on the internal pipe wall or at low spots along the pipe route. On a long-distance pipeline, even a thin film of water will amount to tonnes of remaining water. If water or moisture is left within the pipeline, it can cause a number of problems including:

- Formation of hydrates in 'sweet' gas pipelines.
- Pitting corrosion caused by 'sweet' corrosion, where carbon dioxide dissolves in the water, forming carbonic acid.
- In 'sour' operating pipelines, where hydrogen sulphide (H_2S) is present, the H_2S dissolves in the water, causing pitting corrosion, sulphide stress cracking and hydrogen pressure induced cracking (see Chapter 5).

The aim of drying, therefore, is to reduce the dewpoint of moisture trapped in the pipeline (i.e. the dryer the pipeline, the lower the dewpoint). Figure 4.5 shows the available drying methods.

4.2.1 Methanol swabbing

As shown in Fig. 4.6, an effective and quick approach is to use a train of pigs containing a methanol solution, followed by a drying gas such as nitrogen. This 'methanol swabbing' method is useful as it prevents the formation of hydrates in the pipeline. The methanol mixes with water, creating a swabbing solution that is then moved along via the trailing pigs. The advantage with this method is that the pipeline can be commissioned in one complete operation.

Methanol swabbing

Vacuum drying

Hot gas drying

Nitrogen and dry air

Figure 4.5   Pipeline drying methods

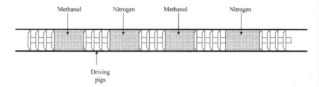

Figure 4.6   Methanol swabbing process

4.2.2 Drying using air or nitrogen along with foam pigs

Nitrogen or air is often used as a drying gas since both gases have a low dewpoint. Foam pigs are then passed through the pipeline, progressively drying it out. Acceptable dryness levels are chosen by measuring the dewpoint of the remaining gas within the pipeline.

4.2.3 Operational drying (applicable to gas pipelines)

By carefully controlling the pressure, nitrogen gas is introduced into the pipeline along with the product. This continues until the dewpoint falls to an acceptable level. Once this is reached, the pipeline pressure is then raised up to the required operational level.

4.2.4 Vacuum drying

Vacuum drying involves reducing the pressure in the pipeline until the remaining water begins to boil off. This pressure is known as the vapour pressure. Figure 4.7 shows the basic principle. Assuming the pipeline is at ambient temperature conditions, a pressure is reached where the water starts to boil off. At atmospheric pressure, this boiling point is at 100 °C. In practice, this consists of three phases:

- *Evacuation*. During this phase, pressure is reduced to a level causing the water to evaporate at the ambient temperature.
- *Evaporation*. Once this pressure is reached, it is maintained to ensure that all the free water is evaporated.
- *Final Drying*. The final process is to remove all water vapour from the pipeline. This is achieved by further

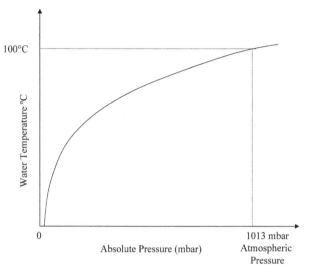

Figure 4.7   Vacuum drying

reducing the pressure using vacuum equipment, drawing the remaining vapour out of the pipeline.

Figure 4.7   Vacuum drying

Chapter 5

Pipeline Operation

The condition of a pipeline must be continually monitored to ensure that defects do not cause it to fail. This is often done through monitoring the pipeline using a combination of approaches:

- aerial surveillance;
- walking survey;
- intelligent inspection tools;
- SCADA system (supervisory control and data acquisition system);
- pipeline cathodic protection and coating surveys.

Typical frequencies of these types of survey are shown in Table 5.1. With the use of risk-based inspection (see Chapter 3), inspection frequencies using intelligent inspection tools will vary between pipelines. Pipeline surveillance involves aerial surveillance using a helicopter to fly over the pipeline route, identifying any areas of activity that could potentially damage the pipeline. Foot patrol surveys are also conducted, to provide more detailed information. Both these survey methods can only tell the operator when there is a risk of third party damage. Damage due to internal or external corrosion is easily monitored through inspection using intelligent tools, or 'pigs'.

Pigs can be divided into two main types:

Table 5.1 Typical frequencies for pipeline monitoring activities

Aerial surveillance	Walking survey	Intelligent inspection	CP system check
2 weeks	4 years	Max. 10 years	Max. 10 years

- cleaning pigs and gauge pigs;
- intelligent inspection pigs (also called 'inspection tools').

The purpose of a gauge pig is to prove the internal diameter of the pipe by checking for dents or partially closed valves prior to launching the intelligent inspection tool. If one of these tools were to get stuck inside a pipeline then it would need to be cut out and the section replaced, causing huge costs for the operator. A gauge pig consists of an aluminium circular plate with a diameter 95% of the internal bore. If the gauge pig emerges with no damage to the plates, the pipeline is considered acceptable for pigging.

Cleaning pigs are often made from foam which cleans the inside of a pipeline from debris such as welding rods and absorbs trapped dirt or soil. Alternatively, brush pigs are used repeatedly throughout the pipeline to ensure that all the debris has been removed (see Fig. 5.1).

The main pig is the intelligent inspection tool, of which

Courtesy of Rosen Inspection

Figure 5.1   Brush cleaning pig

there are many different types. Currently there is no inspection tool available that will detect every type of defect. The main technologies at present include:

- magnetic inspection;
- transverse field inspection;
- ultrasonic inspection;
- ultrasonic crack detection;
- electromagnetic inspection;
- geometric inspection tools.

Inspection tools are able to detect numerous types of defect. These are summarized in Table 5.2. Other specialist tools exist which can provide an accurate three-dimensional route mapping of the pipeline along with global positioning readings. The most commonly used tools are the magnetic and ultrasonic inspection tools.

5.1 Magnetic inspection

Magnetic inspection involves saturating the pipe wall with an axial magnetic field using magnetized brushes in contact with the pipe wall (see Fig. 5.2). The sequence of operation is as follows:

- The tool is pushed along the pipeline using the flow of the product. The gas or oil creates a driving force on the rubber cups, which forces the inspection tool along the pipeline. Generally, the inspection tool velocity does not exceed ~4 m/s to ensure that the quality of the data received is within acceptable limits.
- As the tool travels through the pipe, powerful magnets saturate the pipe wall, using brushes in contact with the internal wall. A sensor ring between the magnets picks up any changes in the magnetic field.
- If a defect is found, this causes a leakage in the magnetic field, which sensors can detect.
- The sensor ring located in the first section of the tool only detects the presence of metal loss and does not know

Table 5.2 Inspection tool capabilities

	Inspection capabilities	Limitations

Magnetic	General corrosion
	Pitting corrosion
	Dents
	Gouges
	Manufacturing defects
	Good at detecting girth
	  weld anomalies such
	  as lack of fill
		(1) Cannot size dents
		(2) Cannot size cracks
		(3) Only sensitive to
		     features orientated
		     in the circumfer-
		     ential direction

Transverse
  field inspection
	General corrosion
	Pitting corrosion
	Seam weld anomalies
	Dents
	Gouges
	Manufacturing defects
		(1) Cannot size dents
		(2) Cannot size cracks
		(3) Only sensitive to
		     features orientated
		     in the axial direction

Ultrasonic
  inspection
	General corrosion
	Pitting corrosion
	Seam weld anomalies
	Dents
	Gouges
	Manufacturing defects
	Laminations
		(1) Requires a liquid
		     medium for
		     operation
		(2) Cannot detect stress
		     corrosion cracking
		(3) Cannot size dents

Ultrasonic
  crack detection
  tools
	Seam weld cracks
	Stress corrosion cracking
	Other planar defects
		(1) Requires a liquid
		     medium
		     for operation
		(2) Limited corrosion
		     detection capability

Electromagnetic
  inspection
	Can be used for
	  inspection of risers
	Stress corrosion cracking
		(1) Relatively more
		     expensive than the
		     other tools

Caliper	Can accurately size
	  dents and geometric
	  distortions in the pipe
	  wall such as ovality
		(1) Can only be used for
		     profile distortions

where it is in the pipe wall. This is done through sensors attached to the trailing section of the tool, that detect whether the defect is on the internal or external surface.

- Other measurements are taken using these tools, such as pipe wall thickness, odometer distance, tool velocity, orientation, pressure and temperature.

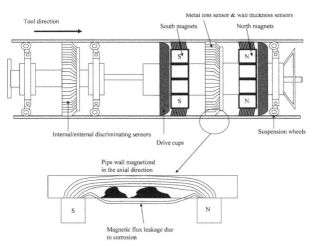

Figure 5.2   Magnetic inspection tools

Data received from intelligent inspection tools are not direct measurements – the magnetic information has to be interpreted by analysts who then discriminate between general and pitting corrosion, dents, gouges and manufacturing defects. Hence, these data have to be accurately interpreted and sized correctly. Data quality can be affected by a number of factors:

- how clean the pipeline is;
- velocity excursions (i.e. if the tool speed is too fast or slow, data may be lost);
- faulty sensors;
- analyst interpretation.

The typical accuracy of a magnetic inspection tool is shown in Table 5.3. More recent developments in the inspection tool industry have seen the development of transverse field inspection. This works on the same principle by magnetizing the pipe wall and detecting field changes. The pipe wall is

Table 5.3   Typical magnetic tool capabilities

Size range	Typically 6–56 inches
Operating product type	Liquid or gas
Operating range	Typically up to 80 km <10″, up to 150 km >10″
Operating temperature	Up to 40 °C
Acceptable bends	Tolerate as low as 3 x pipe diameter
Location accuracy	Axially, ± 100 mm from welds Circumferentially, ± 5°
Depth accuracy	Typically ± 10% wall thickness

magnetized in the circumferential direction, which means that these tools are capable of detecting axially oriented features with much better accuracy. This tool is useful for detecting channelling internal corrosion due to water drop-out, or in pipelines that have long seam weld defects.

5.2 Ultrasonic inspection

These tools are often used to conduct a baseline survey to view the condition of a newly constructed pipeline. Unlike magnetic tools, the ultrasonic inspection tool is able to make direct measurements of defects and the remaining wall thickness. This is done by the use of ultrasonic technology, where sound waves are emitted through the thickness of the pipe. Sound waves travel through the steel and are reflected at the outer wall, and their return is detected using sensors (see Fig. 5.3). This is known as the 'echo time' and is used to determine feature dimensions on the basis of the time it takes to return back to a sensor. A benefit of this is that mid-wall features such as laminations and blisters can easily be detected using crack detection tools.

Ultrasonic inspection usually requires a liquid medium for it to operate. Consequently, this would only be used in oil

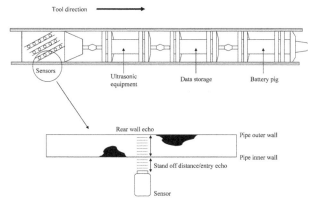

Figure 5.3   Ultrasonic inspection tools

pipelines and cannot be easily applied in gas pipelines. The sequence of operation is as follows:

- Sensors are set a small distance from the inner wall of the pipe, known as the 'stand-off' distance.
- Sound waves are emitted perpendicular to the pipe wall from a transmitter. The transmitter triggers a pulse across the stand-off distance, as shown in Fig. 5.3. This creates an entry echo (as the tool passes an internal metal loss defect, this distance will increase).
- The remaining sound waves pass through the steel and are reflected at the outer wall. This is known as the 'rear wall echo' and can be used to measure the wall thickness. If the 'entry echo' remains unchanged, any subsequent changes in wall thickness suggest that metal loss is external.
- Consequently, the value for wall thickness is represented by the distance between the entry echo and the rear wall echo.

The typical accuracy of ultrasonic inspection tools is shown in Table 5.4. More recent developments include crack detection tools specially designed to detect features such as

Table 5.4 Typical ultrasonic tool capabilities

Size range	Typically 6–60″
Operating product type	Liquid (specialist tools are required for gas operation)
Operating range	Typically up to 1000 km
Operating temperature	Up to 50 °C
Acceptable bends	Tolerate as low as 1.5 x pipe diameter
Location accuracy	Axially, +200 mm from welds
	Circumferentially, dependent on tool sensor spacing
Depth accuracy	Typically ±0.5 mm

stress corrosion cracking or fatigue cracks (see Fig. 5.4). Instead of transmitting sound waves perpendicular to the pipe wall, pulses are transmitted at 45° circumferentially through the pipe wall, which enables fine cracks to be detected. Again, measurements such as odometer distance, tool velocity, orientation, pressure and temperature are taken.

5.3 Geometric tools

Magnetic or ultrasonic metal loss tools can detect the presence of geometric changes in curvature of the pipe wall, but they cannot be used to size them directly. As a result, specialist geometric tools are required that can accurately determine the depth and profile of features such as ovality, dents and buckling. These are often used prior to launching intelligent inspection tools, to check the pipeline for dents or other deformations that could potentially block the tool (see Fig. 5.5).

The inspection of offshore sections of pipe such as risers is more complicated since conventional intelligent tools require a rigid pipe body. Flexible risers are constructed of different

Courtesy of Rosen Inspection

Figure 5.4   Rosen inspection crack detection tool

layers of steel (i.e. a metal composite). These types of pipe are often inspected using eddy current pigs, which operate by inducing an eddy current in the metal composite and detecting the difference in electromagnetic field caused by features such as cracks.

As this chapter has shown, there is no one tool that can detect every type of feature, and hence it is important to choose the most appropriate inspection tool depending on the history of the pipeline and the types of defect feature that are expected.

5.4 In-service defects and corrosion mechanisms

In understanding defect formation and their failure mechanisms, it is important to consider the different types of defect that may appear. Pipe defects are features that affect the structural integrity of the pipeline, and may be located on the

Courtesy of Rosen Inspection

Figure 5.5   Geometric inspection tool

surface of the pipe wall or inside the material of the pipe. Possible sources of damage include:

- manufacturing;
- construction;
- third party interference;
- operational damage;
- other causes (i.e. ground movement or seismic activity).

5.4.1 Manufacturing

Manufacturing features often take the form of a discontinuity in the geometry of the pipe, such as a reduction in wall thickness or defect in the material itself. These include porosity, inclusions, surface laps or more serious types of manufacturing feature such as laminations, which can bulge or even burst the pipe (especially in sour operating environments). Manufacturing features are often found in older seamless pipes and are mainly due to the nature of the manufacturing process which was previously not as controlled as newer manufacturing methods (e.g. ERW, LSAW, SSAW).

5.4.2 Construction

Construction defects may include girth or seam weld defects caused by lack of fill, misalignment or, in the most severe cases cracking. Other forms of damage may occur such as indentation damage, corrosion at the girth welds, or even damage to the external pipe coating. These generally occur during the construction stage where the pipe spools are positioned at the side of the trench, welded together and then positioned in the trench. Sometimes indentations at the bottom of the pipe may occur due to rocks at the bottom of the trench. These are termed constrained dents, and are not necessarily considered serious, as they are not susceptible to being made worse by fatigue.

5.4.3 Third party interference

Third party damage is often the most severe form of damage, resulting in failure of the pipe or requiring immediate repair. Often this involves mechanical damage such as a gouge resulting in metal loss of the pipe wall, or distortion of the pipe wall such as a dent. A combination of both a dent and a gouge will significantly lower the burst strength of a pipe. When describing dents, these can be broadly categorized into the following:

- *Smooth*. Dents that have a smooth profile and no change in wall thickness.
- *Plain*. Dents that contain no stress-raising features such as corrosion.
- *Constrained*. Dents that are not free to move under the influence of pressure.
- *Unconstrained*. Dents that are free to move under the influence of pressure.

5.4.4 Operational damage

Defects arising from operational usage are mainly corrosion based, i.e. external corrosion caused by damaged or disbonded coating where the CP system is not effective or internal corrosion caused by water in the product.

5.4.5 External corrosion

Costs due to corrosion increase over the lifetime of a pipeline until it is no longer cost effective to continue to use it and a new pipeline is required. As shown in Fig. 5.6 for pipeline A, these costs are attributed to integrity and maintenance activities such as pipeline repairs and corrosion monitoring/ prevention activities.

External corrosion defects formed during operational usage are often caused by damaged or disbonded coating. Corrosion also occurs when the CP system is not effective. As shown in Chapter 2, the primary means of protection against external corrosion is the pipeline coating. The CP system should only be used as a secondary protective measure, to prevent corrosion occurring at defects. However, coating failures can and do often occur, such as:

- surface contamination;
- soil stress cracking;
- high-temperature coating damage;
- disbonding/loss of adhesion.

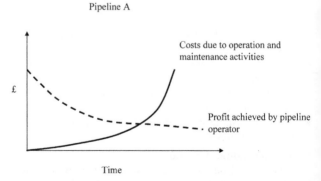

Figure 5.6   Corrosion costs

Applying this to pipelines, the following is a summary of common problems often associated with coating methods:

Field applied tapes

- Linear areas of corrosion associated with a seam weld (caused by tenting or wrinkling of the coating over the weld).
- On spirally welded pipe, areas of spiral corrosion can occur owing to poor tape overlap.
- Concentrated areas of corrosion next to a girth weld indicate that the field applied tape has been incorrectly applied over the weld.
- Moisture or other surface contaminants can cause bubbling during application of coatings.
- Thermoplastic tape wraps and coal tar enamel coatings may be prone to creep at high temperatures.
- Rocks can easily damage cold applied laminate tapes during construction.

Coal tar enamel coatings

- Corrosion occurring at the 6 o'clock position would suggest that this is likely to have been damaged during construction.
- Linear corrosion occurring at the 12 o'clock position would suggest that soil stresses could have damaged the coating. Since pipelines are covered in soil, sometimes the overburden pressure can cause the coating to sag or crack over the top half of the pipeline. Coal tar enamel coatings are prone to this type of failure.
- Coating disbonding can occur downstream of pressure reduction stations, where low temperatures are present.

What causes corrosion, can it be easily detected and what should you do? As discussed in Chapter 3, for corrosion to occur, two areas of different potential need to exist, i.e. an anode and a cathode.

When a potential difference occurs between the anode and cathode, and they are connected through the electrolyte (i.e. surrounding soil), then conditions exist for a corrosion cell to occur. During this process, iron is oxidized (known as the anodic reaction) which releases electrons. The resulting free electrons are consumed at the cathode in a reduction reaction forming hydroxide, as shown below:

$$Fe \rightarrow Fe^{++} + 2e^- \text{ (anodic reaction)}$$
$$Fe^{++} + 2OH^- \rightarrow Fe(OH)_2$$
$$O_2 + 2H_2O + 4e^- \rightarrow 4OH^- \text{(cathodic reaction)}$$
$$2H_2O + 2e^- \rightarrow H_2 + 2OH^- \text{(cathodic reaction)}$$

Figure 5.7 shows an example of corrosion occurring under disbonded coating. The main types of corrosion are general corrosion, pitting corrosion and stress corrosion cracking. Pitting corrosion shows as a very localized area of attack, and as a result this type of corrosion is often deep, isolated and growing in a vertical direction relative to the pipe

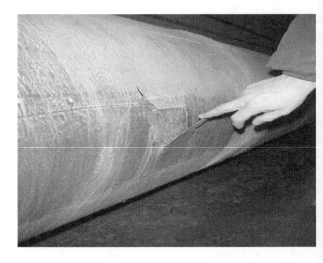

Figure 5.7   Corrosion feature

surface. Areas of general corrosion tend to be shallower and are randomly spread over a larger surface area.

Another more serious type of corrosion is bacterial corrosion or microbiologically induced corrosion (MIC) which can occur in organic soils such as river clays. These soils often contain sulphate-reducing bacteria (SRB), which produce very localized areas of accelerated corrosion growth. Currently, the best tools to detect these types of feature are the magnetic and ultrasonic inspection tools. By its nature, corrosion is random and is sensitive to different soil types and pipe properties such as:

- moisture content;
- oxygen levels;
- pH levels;
- pipe grade.

One of the most severe forms of corrosion is stress corrosion cracking (SCC), which occurs under conditions of an applied tensile stress and corrosive environment. This process forms small cracks that are perpendicular to the applied stress. Pipeline steel can be susceptible to this form of damage under repeated pressure cycling. Finally, if corrosion is detected on a pipeline, the corrosion growth rates will be different along the length of the pipeline. Corrosion analysis is a complex process and involves the types of key decision shown in Fig. 5.8.

5.4.6 Internal corrosion
Internal corrosion occurring during operational conditions is caused by water or moisture trapped in the product. Again, this can be in the form of pitting or general corrosion, but the two main mechanisms are:

- sweet corrosion;
- sour corrosion.

These forms of corrosion are present in oil or gas pipelines and are dependent on the content of hydrogen sulphide and

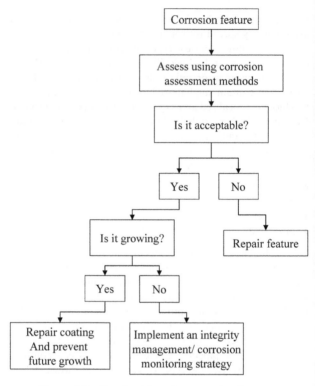

Figure 5.8 Key decisions for corrosion features

carbon dioxide in the pipeline. NACE define a partial pressure of 0.05 psia, above which it is termed 'sour' operating conditions [**14**].

5.4.6.1 Sweet corrosion

Conditions for sweet corrosion are shown in Fig. 5.9. For sweet corrosion to occur, the pipeline must contain carbon dioxide and only small levels of hydrogen sulphide (i.e. a partial pressure *below* 0.05 psia hydrogen sulphide). During this process, carbon dioxide dissolves in free water to form carbonic acid, which corrodes the pipe wall. As the concentration of carbon dioxide increases, so does the corrosion rate. This type of corrosion tends to form areas of general and pitting corrosion.

5.4.6.2 Sour corrosion

As the concentration levels of hydrogen sulphide increase, 'sour' operating conditions start to prevail (i.e. a partial pressure of 0.05 psia and *above* of hydrogen sulphide). Under these operating conditions the predominant failure mechanism is hydrogen cracking, of which there are several types:

- *Hydrogen-induced cracking (HIC)*. HIC is associated with blistering of the pipe and is also commonly called hydrogen pressure induced cracking. During operation in sour conditions, hydrogen sulphide reacts with the pipeline steel to form a thin film of iron sulphide. Under these conditions, as corrosion occurs, atomic hydrogen diffuses into the pipeline steel and recombines to form hydrogen gas at discontinuities in the microstructure.

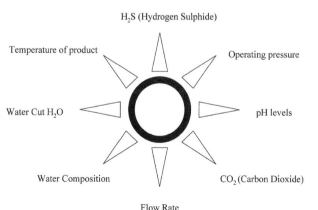

Figure 5.9 Conditions for sweet corrosion

69

These discontinuities are usually areas of manganese sulphide inclusions formed during the manufacturing process or lamination features. Finally, as hydrogen gas begins to build up in these areas, this increases the local stresses, causing the pipe to bulge and form blisters. Cracks then form and propagate through the pipe in the form of stepwise cracking.

- *Sulphide stress corrosion cracking (SSCC)*. This failure mechanism is by hydrogen embrittlement and forms in a similar way to hydrogen-induced cracking. Atomic hydrogen forms a solid solution in the steel microstructure, reducing the ductility of the material, i.e. hydrogen embrittlement occurs. Cracking then takes place under conditions of an applied tensile stress and propagates under fatigue loading conditions. Higher-grade steels are particularly susceptible to this form of damage. In summary:
  - SSCC is influenced by operating stress.
  - HIC is based on material properties.
  - the formation of cracks and failure can be very rapid.
- *Stress-oriented hydrogen-induced cracking (SOHIC)*. This form of damage occurs as a combination of HIC and SSCC. Here, stepwise cracks form in areas of high stress caused by build-up of hydrogen (HIC) and stress cracking occurs owing to embrittlement in these areas.

The following section details common patterns of internal corrosion.

5.4.7 Gas pipelines

Figure 5.10 shows an example of corrosion concentrated at the start of a pipeline and concentrated at the bottom half of the pipeline. This type of corrosion pattern would be indicative of high levels of water within the product, resulting in water drop-out at the start of the pipeline. Condensation of water is temperature sensitive and can often result in a corrosion peak further along the pipeline route, as the temperature of the gas decreases.

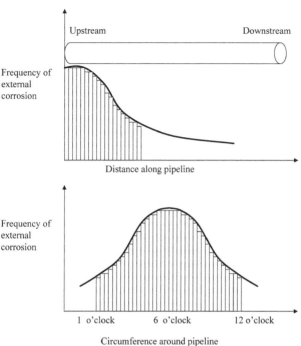

Figure 5.10 Typical corrosion distribution

Consideration should also be given to:

- preferential corrosion associated with a girth weld, which can be caused by beaded edges at the girth weld that disrupt flow, creating lines of condensation;
- low spots along the pipeline route, where water condensate can accumulate;
- corrosion peaks associated with temperature changes (i.e. where the gas has cooled sufficiently to allow condensation);
- pre-service corrosion patterns, such as where pipeline spools are not stored with end caps.

5.4.8 Liquid pipelines

Like gas pipelines, the main problem is associated with water drop-out from the product, causing distinctive corrosion patterns. In particular, long areas of 'channelling' corrosion at the bottom of the pipe are associated with water trapped at the bottom of the pipe. Generally, water can separate from the oil phase at the start or further along the pipeline route, creating corrosion peaks. Consideration should therefore be given to:

- low spots along the route;
- corrosion peaks associated with temperature changes (where water comes out of solution at a critical temperature);
- channelling corrosion patterns.

5.4.9 Protection from internal corrosion

An effective way of preventing internal corrosion is the application of corrosion inhibitors within the product or applying an internal coating through the pipeline. A limitation of inhibitors is that they are sensitive to flow velocity: if flow velocity is too high, then the inhibitor will not effectively protect the pipe wall.

Corrosion should be assessed as part of an overall fitness-for-purpose assessment, to determine:

- why the corrosion has occurred;
- whether the corrosion is currently active;
- whether it affects the immediate integrity of the pipeline;
- which assessment approach should be used.

Finally, corrosion should be continually monitored through internal inspections. Following these internal inspections, pipeline operators often utilize fitness-for-purpose (FFP) assessments to assess the acceptability of defects found in the pipe. The following chapter discusses assessment methods commonly used throughout the pipeline industry to determine the acceptability of defects.

Chapter 6

Pipeline Maintenance

Optimizing maintenance costs for pipeline operators is important since the aim is to prevent failures, maintain high standards of pipeline integrity and maximize safety. While it is widely understood within the pipeline industry that the most common cause of pipeline failure is third party damage, pipeline failures can and do often occur as a result of defects in the pipe.

A defect is defined as something that *affects the structural integrity* of a pipeline, and can have a number of causes. Defect assessment is only part of the picture for an overall pipeline integrity and management strategy that aims safely and cost effectively to extend the operating life a pipeline. An integrity and management strategy involves pipeline inspection, risk assessment, defect assessment, operation and maintenance and, finally, repair. There are numerous codes that give guidance on the assessment or classification of defects on oil and gas pipelines. Table 6.1 shows just a few of them. Defects to be considered include:

- stress corrosion cracking;
- internal corrosion;
- external corrosion;
- gouge;
- dent;
- manufacturing defect;
- hydrogen-induced cracking;

Table 6.1　Commonly used pipeline codes

ASME B31.4	ASME B31.G
ASME B31.8	Modified B31.G (RSTRENG)
BS 7910	DNV
API 579	API 1156
EPRG	PD 8010

- girth weld defect;
- seam weld defect.

Unfortunately, there are currently no inspection tools that will detect all of these types of feature. It is therefore important to identify the most likely defects on the basis of information about the pipeline, in particular:

- the age of the pipeline;
- whether there has been a previous inspection or baseline inspection;
- operational conditions (i.e. sour or sweet conditions);
- whether the pipeline is onshore or offshore;
- the type of pipe/manufacturing method (certain types of pipe are more prone to certain defects; in particular, older seamless pipe can be susceptible to inclusions and laminations).

Finally, when conducting a detailed assessment of corrosion features, the two main components of stress described in all pipeline codes are:

Hoop stress

$$\sigma_{hoop} = \frac{PD}{2t}$$

Axial stress

$$\sigma_{axial} = \frac{PD}{4t}$$

When assessing the effects of other loads such as high-temperature operation, ground movement or seismic activity, then detailed modelling of the loads may be required. In these cases, the use of finite element analysis may be appropriate.

6.1 Assessment of internal and external corrosion features

It is important to understand the main *failure mechanisms* caused by corrosion. The two main mechanisms are:

- *Leakage*. This results in a relatively small loss of product.
- *Rupture*. Larger defects fail with a sudden release of pressure that can cause propagating or 'running' fractures in isolated cases.

Owing to the plasticity of steel, the majority of pipeline failures result in leakage rather than rupture. However, in some cases, owing to material properties, the steel can rupture in a brittle mode causing long running fractures. There are a number codes available that provide guidance on the assessment of corrosion features before failure, all of which can be divided into two main approaches: the older 'effective area' methods and the newer developed 'UTS-based' methods. The effective area methods assume that the strength loss due to corrosion is proportional to the length of metal loss axially along the pipe. The UTS-based methods assume that the ultimate tensile strength of the material controls propagation of a corrosion defect.

Figure 6.1 summarizes the most commonly used methods, i.e. the effective area methods, which include B31.G [15] and

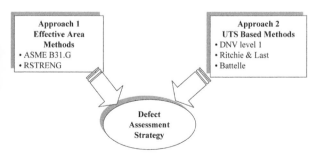

Figure 6.1 Commonly used corrosion assessment methods

simplified RSTRENG (modified B31.G) [16], and those that are UTS based, including DNV-RP-F101 [17], Ritchie and Last (Shell 92 criteria) [18] and PCORRC [19].

Both the effective area and UTS-based methods assume that the corrosion defects are blunt and that the material fails by plastic collapse. The assumption is that any corrosion defect will fail when the stress in the pipe reaches the 'flow stress' of the material. The flow stress is actually a value between the yield strength and ultimate tensile strength of the material (if a corrosion defect were pressurized until failure, the defect would not fail at the yield strength of the material).

Owing to the geometry and plasticity properties of steel, as the defect is just on the point of starting to fail, the region around the defect begins to bulge and deform. This deformation has the effect of work hardening the steel, increasing the stress concentration around the affected region. In addition, the Battelle Institute conducted a great deal of research into the 'folias bulging factor' which takes into account the increased stress concentration as the defect fails.

Two important terms in corrosion assessment equations are therefore:

- the folias factor or bulging factor, which takes into account the work-hardening effect as the corrosion defect begins to bulge and eventually fail;
- the flow stress (i.e. the stress at which the defect is predicted to fail).

Both these terms vary slightly between codes. Important conclusions about the older effective area methods are as follows:

- The methods were developed and validated against older pipeline steels.
- Their approaches could be overly conservative.
- They have limited application to higher-grade steels, i.e.

neither B31G nor RSTRENG is applicable to pipeline steels above grade X65.

Accurate information is needed for any corrosion assessment. The following should be determined:

- location class;
- wall thickness;
- pipeline diameter (nominal);
- pipe grade;
- design pressure or MAOP;
- corrosion dimensions.

Once this information is available, an accurate assessment of corrosion features can be conducted.

6.1.1 Effective area methods

Initial development of the ASME B31.G code [15] occurred in the late 1960s and was based on experimental data from full-size tested pipe sections, using corroded pipe sections and pressurizing them to failure. This enabled a better understanding of defect behaviour, allowing semi-empirical mathematical expressions to be developed and validated against experimental data. The expressions assumed that the failure of blunt corrosion defects is controlled by the yield stress of the pipe material.

The term *effective area* is based on the amount of metal loss on the pipe and assumes that strength loss due to corrosion is proportional to the axial length of the corrosion along the pipe. The basic expression for effective area is as follows:

$$\frac{\sigma_f}{\overline{\sigma}} = \left[\frac{1 - X}{1 - X/M} \right]$$

where; $\overline{\sigma}$ is the flow stress, σ_f is the failure stress, $X = d/t$ or A/A_o, d is the peak depth, t is the wall thickness, A is the area of corrosion, A_o is the original area of the pipe wall and M is the folias factor.

Since steel has a certain amount of plasticity, the material deforms in a way that creates a bulge as the corrosion feature begins to fail. This folias factor M is commonly known as the bulging factor. Figure 6.2 shows the basic idea behind this assessment method. Consequently

$$\frac{A}{A_o} = \frac{d_{peak} L_{overall}}{L_{overall} t} = \frac{d_{peak}}{t}$$

where A is the corroded area, A_o is the original area, $L_{overall}$ is the total length of the corroded area, d is the peak depth and t is the wall thickness.

6.1.2 ASME B31.G assessment method

When applying the effective area method using B31.G, the primary assumptions are as follows:

- The corroded area is approximated as a parabolic shape.

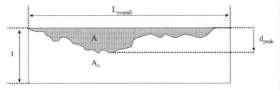

The corroded area A is calculated as shown below;

$A = d_{peak} \times L_{overall}$

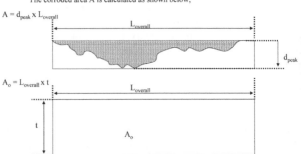

$A_o = L_{overall} \times t$

Figure 6.2   Effective area of corrosion

- The flow stress is based on the yield strength (SMYS) of the steel and is taken as (1.1 × SMYS).
- The peak depth of the defect is less than 80% of the nominal wall thickness.
- The folias factor is calculated using

$$M = \sqrt{1 + 0.893\left(L_m/\sqrt{Dt}\right)^2}.$$

Figure 6.3 shows the shape of the corrosion approximation using B31.G. As can be seen, B31.G approximates the defect as a parabolic shape defect, with an area equal to 2/3 depth × length:

$$\frac{\sigma_f}{\overline{\sigma}} = \left[\frac{1 - (2d)/(3t)}{1 - (2d)/(3tM)}\right]$$

where $\overline{\sigma} = 1.1 \times$ SMYS, σ_f is the failure stress, t is the wall thickness, d is the peak depth and $M = \sqrt{1 + 0.893\left(L_m/\sqrt{Dt}\right)^2}$ (L_m is the maximum length of corrosion and D is the outside diameter).

B31.G was one of the early developed assessment methods and is still one of the most widely used by pipeline operators.

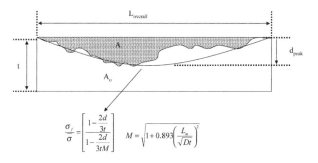

$$\frac{\sigma_f}{\sigma} = \left[\frac{1 - \dfrac{2d}{3t}}{1 - \dfrac{2d}{3tM}}\right] \qquad M = \sqrt{1 + 0.893\left(\frac{L_m}{\sqrt{Dt}}\right)^2}$$

Figure 6.3 B31.G corrosion shape approximation

It was subsequently revised as the modified B31.G [16] criteria or 'simplified RSTRENG' remaining strength assessment method.

B31.G also provides an alternative approach for which a maximum acceptable length for a corroded area can be calculated. This is done using the following equation:

$$L = 1.12B\sqrt{Dt}$$

where D is the outside diameter and t is the nominal wall thickness. Term B is then calculated as

$$B = \sqrt{\left(\frac{d/t}{1.1d/t - 0.15}\right)^2 - 1}$$

Note that B cannot exceed a value of 4 (as shown in Fig. 6.4), so, if the corrosion depth is between 10 and 17.5%, use $B = 4.0$.

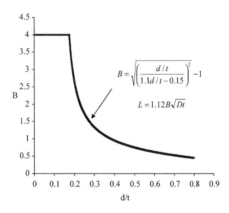

Figure 6.4   Term B for calculating maximum acceptable corrosion length

6.1.3 Simplified RSTRENG assessment method

In order to reduce some of the conservatism found with the original B31.G method, it was modified in the late 1980s, resulting in the simplified RSTRENG criteria [16]. This new method was validated against numerous burst tests on actual corrosion pipe defects. Differences between this new method and the previous one were as follows:

- Rather than approximating the defect as a parabolic shape, the area is approximated as 85% of the peak depth, using a factor of 0.85 (see Fig. 6.5).
- A more accurate three-term expression for the folias factor is provided.

If $L^2/Dt \leqslant 50$, then

$$M = \left[1 + 0.6275\left(\frac{L^2}{Dt}\right) - 0.003375\left(\frac{L^2}{Dt}\right)^2\right]^{1/2}$$

If $L^2/Dt > 50$, then

$$M = 0.032\left(\frac{L^2}{Dt}\right) + 3.3$$

where L is the length of the corroded area, D is the outside diameter and t is the nominal wall thickness.

A less conservative flow stress is also provided (i.e. the stress before failure is slightly higher than that used in B31.G):

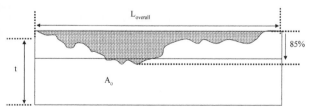

Figure 6.5 Corrosion shape approximation

$$\bar{\sigma} = \text{SMYS} + 10\,\text{ksi} \quad \text{or} \quad \text{SMYS} + 68.95\,\text{MPa}$$

Applying these changes, the equation for the modified B31.G criteria becomes

$$\frac{\sigma_f}{\bar{\sigma}} = \left[\frac{1 - 0.85(d/t)}{1 - 0.85(d/t)(1/M)}\right]$$

In summary, the simplified RSTRENG is *less conservative* than B31.G because:

- It includes a more accurate three-term expression for the folias factor (or bulging factor).
- It contains a less conservative prediction for flow stress.
- Only 85% of the peak depth is used in its assessment. This method is more suitable for long areas of general corrosion.

The citation to this reference (reference [16]) conveys no rights to the reader in the material referenced and it may not be used without the prior written permission of Pipeline Research Council International, Inc.

6.1.4 UTS-based methods

Unlike the older effective area methods, more recently developed corrosion assessment methods have assumed that failure is controlled by the ultimate tensile strength of the material. These approaches include DNV [17], Ritchie and Last [18] and the Battelle PCORRC assessment criteria [19].

6.1.5 DNV-RP-F101

Unlike B31.G and RSTRENG, the DNV assessment approach takes into account other loading conditions such as compressive axial loads. DNV-RP-F101 is a result of a joint industry project between DNV [17] and BG Technology. Both BG Technology and DNV generated an extensive database of burst tests on pipe specimens, incorporating single corrosion defects, interacting defects and complex corrosion shapes. DNV generated a database of 12 burst tests and BG generated a database of more than 70 burst tests [17]. Alongside these experimental data, a number of three-dimensional non-linear finite element pipe models containing defects were produced and validated against experimental data. This then enabled criteria to predict the remaining strength of corrosion defects to be developed.

DNV-RP-F101 is divided into two main assessment approaches for assessing single, interacting and complex-shaped corrosion. Part A is based on a 'calibration approach' taking into account uncertainties of different inspection tool accuracy, so a number of safety factors are included in the equation for calculating failure pressure. Part B employs the same equation for failure pressure but takes a simpler approach using an 'allowable stress design' format. Corrosion failure pressure is then multiplied by a safety factor to obtain the safe working pressure, and hence uncertainties in sizing are left to the user to take into account.

Considering a single corrosion defect for this assessment approach, the failure pressure of the corrosion defect is first calculated and then multiplied by a safety factor, which includes a combination of a 'modelling factor' and the pipeline 'design factor'. This safety factor is known as the 'total usage factor' and is based on the following:

Total usage factor F is $F_1 \times F_2$

where $F_1 = 0.9$ is the modelling factor (based on a 95% confidence interval) and F_2 is the operational usage factor. This is introduced so that there is a safety margin between the operating pressure and failure pressure of the corrosion defect (this is normally taken to equal the design factor of the pipeline).

There are therefore two stages in the assessment approach:

- Calculate the failure pressure of the corrosion defect:

$$P_{\mathrm{f}} = \frac{2tf_{\mathrm{u}}}{D-t} \frac{(1-d/t)}{(1-d/(tQ))}$$

where $Q = \sqrt{1 + 0.31\left(l/\sqrt{Dt}\right)^2}$, f_{u} is the ultimate tensile strength, D is the outside diameter, t is the nominal wall thickness and P_f is the failure pressure.

- Then calculate the safe operating pressure using the total usage factor:

$$P_{swp} = FP_f$$

If the peak depth of the corroded area is greater than 85% of the wall thickness (as opposed to 80% for B31.G), then this is considered unacceptable. In summary, the main differences between this assessment approach and B31.G and simplified RSTRENG are as follows:

- The maximum allowable depth is 85% of the wall thickness.
- It assumes a rectangular flat-bottomed defect.
- Failure is controlled by the UTS (the ultimate tensile strength of the material).
- It utilizes a different folias factor:

$$Q = \sqrt{1 + 0.31 \left(\frac{l}{\sqrt{Dt}}\right)^2}$$

6.1.6 Ritchie and Last (Shell 92 criteria)

In response to the conservatism associated with B31.G, these criteria were developed and are also known as the Shell 92 criteria [18]. Like DNV, this method simplifies the shape of the corrosion defect as a flat-bottomed rectangular defect, and assumes that failure is dependent on the ultimate tensile properties of the material. The difference, however, is that the Shell 92 criteria use only 90% of the ultimate tensile strength:

$$\sigma_f = 0.9(UTS) \left[\frac{1 - d/t}{1 - d/(tM)}\right]$$

where $M = \sqrt{\left[1 + 0.8(L)^2/(Dt)\right]}$, d is the peak depth, t is the wall thickness, D is the outside diameter and L is the length of the corroded area.

As shown in the above equation for calculating the failure stress, this uses the same effective are equation.

The citation to this reference (reference [18]) conveys no rights to the reader in the material referenced and it may not be used without the prior written permission of Pipeline Research Council International, Inc.

6.1.7 Battelle PCORRC assessment criteria

Research conducted by the Battelle Institute [18, 19] suggested that there was more than one failure mechanism of corrosion defects, depending on whether they were located in high- or low-toughness pipe material. These failure mechanisms were as follows:

- Defects in moderate- to high-toughness pipe fail by *plastic collapse* and the UTS controls this failure.
- Defects in low-toughness pipe (such as in older pipeline steels) fail in a *toughness-dependent mode*.

As a result of this research, new residual strength criteria were developed for defects in moderate- to high-toughness pipe. As part of this investigation into failure of corrosion in a plastic collapse mode, Battelle developed special-purpose finite element software in order better to understand some of the controlling factors of plastic collapse. These criteria are only applicable to pipe operating above the brittle–ductile transition temperature (i.e. the failure mode would be plastic collapse) and to pipe whose Charpy upper shelf energy is 61 J and above [18, 19]. As a result, the following equation, known as PCORRC, was developed for defects in moderate- to high-toughness pipe:

$$\sigma_{\mathrm{f}} = \mathrm{UTS}\left\{1 - \frac{d}{t}\left[1 - \exp\left(-0.157\frac{L}{\sqrt{R(t-d)}}\right)\right]\right\}$$

The citation to these references (references [18] and [19]) conveys no rights to the reader in the material referenced and it may not be used without the prior written permission of Pipeline Research Council International, Inc.

6.1.8 Application of these methods in practice and definition of safety factors

In practice, when data are available on corrosion, assessment will be based on:

- data from an intelligent inspection tool;
- actual measurements from on-site investigations.

The benefit of using inspection tools is that the operator can get a clearer view of the number and distribution of defects along the pipe. Often when data are received from an ultrasonic or magnetic inspection tool, there will be hundreds or even thousands of reported internal and external corrosion features, particularly in some of the older pipelines. Assessment curves are therefore an effective method to show whether these features are acceptable at the current maximum allowable operating pressure (MAOP) of the pipeline.

Figure 6.6 shows peak corrosion depth against axial length. Using assessment codes, the operator can calculate failure depths for different lengths of corrosion at the current MAOP.

It is important for any engineer when conducting calculations to take into account uncertainties and operational considerations. These may include:

- uncertainties in defect measurement (inspection companies usually provide conservative estimates for peak depth);
- inconsistencies in material properties (changes in wall thickness or yield strength);
- density of population surrounding the pipeline (codes such as B31.8 provide different location classes);
- maximum operating pressure of the pipeline (traditionally, pipelines are operated to a maximum 72% SMYS).
- overpressures (operators should take into account the

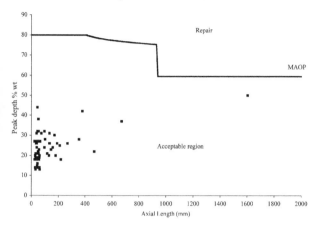

Figure 6.6   Plot of peak depth versus axial length

effect of surge or hydrostatic effects due to elevation changes).

B31.G [15] includes such a safety margin by defining the measure of acceptance as follows: corrosion defects that can survive a hydrotest to 100% SMYS will be acceptable for operation to 72% SMYS, therefore resulting in a safety factor of 1.39 (100/72 = 1.39). Hence, a number of safety factors (SFs) for different location classes can be defined using the following equation:

$$SF = \frac{1}{DF}$$

where DF is the design factor.

Employing the safety factors in Table 6.2 for different location classes 1 to 4, this can be used to calculate whether a corrosion defect is acceptable. For example, applying this to Fig. 6.6 for a gas pipeline located in a location class 1 area, it is possible to show which features would fail by applying a safety factor of 1.39 to the MAOP.

Table 6.2 Calculation of safety factors

Location class	Design factor	Calculation	Safety factor
1	0.72	1/0.72	1.39
2	0.60	1/0.6	1.67
3	0.50	1/0.5	2.00
4	0.40	1/0.4	2.50

Figure 6.7 shows that one corrosion feature is above the repair line at 1.39 × MAOP, meaning that this feature should be repaired. It is possible to apply this procedure for any of the codes shown in Table 6.3 by simply plotting the acceptance curves (note that RSTRENG allows a safety margin of 1.39 × MAOP for any location classification [16]). In addition, when calculating safety factors for DNV, the total usage factor should be taken into account ($F = 0.9 ×$ design factor), for example:

$0.72 × 0.9 = 0.648$

therefore $1/0.648 = 1.54$ for location class 1.

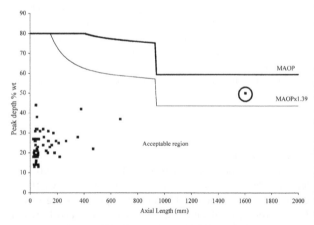

Figure 6.7 Application of safety factor

Table 6.3   Safety factors for different location classifications

Assessment code	Location class	Design factor	Safety factor
B31.G	1	0.72	1.39
	2	0.60	1.67
	3	0.50	2.00
	4	0.40	2.50
RSTRENG	1	0.72	1.39
	2	0.60	1.39
	3	0.50	1.39
	4	0.40	1.39
DNV	1	0.72	1.54
	2	0.60	1.85
	3	0.50	2.22
	4	0.40	2.78

Finally, it is important to choose the most appropriate inspection tool for the pipeline. Figure 6.8 shows a summary of the basic information that most inspection tools should provide.

In developing a corrosion management strategy it is essential to determine if corrosion growth is occurring along the pipeline route. This can only be achieved through repeat inspections or the use of corrosion coupons placed alongside the pipe. The benefits of understanding corrosion rates are that they allow the operator to:

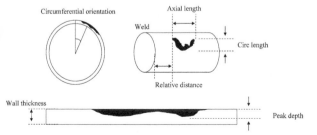

Figure 6.8   Basic information provided by inspection tools

- prioritize repairs;
- set re-inspection intervals;
- predict future repair dates.

6.1.9 Example 1: assessment of stress

A 24 in diameter pipeline with a wall thickness of 9.5 mm is to be operated at 70 bar. Assuming no other secondary loads, calculate the stress components acting on the pipeline:

Hoop

Since

$$\sigma_{hoop} = \frac{PD}{2t}$$

it follows that

$$\sigma_{hoop} = \frac{7.0 \times 609.6}{2 \times 9.5} = 224.59 \, \text{N/mm}^2$$

Axial

Since

$$\sigma_{axial} = \frac{PD}{4t}$$

it follows that

$$\sigma_{axial} = \frac{7.0x609.6}{4x9.5} = 112.29 \, \text{N/mm}^2$$

These results show that the axial stress is half the hoop stress component.

6.1.10 Example 2: assessment of axial dimensions

A 24 in crude oil pipeline has been inspected using an ultrasonic inspection tool. Results show that the majority of the reported corrosion features are external. In addition, ultrasonic measurements show that the minimum wall thickness is 8.7 mm, and the deepest reported corrosion feature is at 55% wall thickness and has an axial length of 200 mm. If the pipeline material is X52 grade, and is operating at an MAOP of 45 bar, calculate the flow stress and bulging factor using:

- B31.G;
- simplified RSTRENG.

6.1.10.1 B31.G – flow stress

Since the flow stress $\overline{\sigma} = 1.1 \times$ SMYS, and SMYS for X52 grade is $358\,\text{N/mm}^2$, it follows that $\overline{\sigma} = 1.1 \times 358 = 393.8\,\text{N/mm}^2$

6.1.10.2 Simplified RSTRENG – flow stress

Since the flow stress $\overline{\sigma} = $ SMYS $+ 68.95\,\text{MPa}$, it follows that $\overline{\sigma} = 358 + 68.95 = 426.95\,\text{N/mm}^2$.

6.1.10.3 B31.G – bulging factor

As defined in ASME B31.G, the folias bulging factor

$$M = \sqrt{1 + 0.893\left(\frac{L_m}{\sqrt{Dt}}\right)^2}$$

Therefore, if the corrosion length is 200 mm, the diameter is 609.6 mm and the wall thickness is 8.7 mm, it follows that

$$M = \sqrt{1 + 0.893\left(\frac{200}{\sqrt{609.6 \times 8.7}}\right)^2} = 2.78$$

6.1.10.4 Simplified RSTRENG – bulging factor

As defined in simplified RSTRENG, if $L^2/Dt \leqslant 50$, then

$$M = \left[1 + 0.6275\left(\frac{L^2}{Dt}\right) - 0.003375\left(\frac{L^2}{Dt}\right)^2\right]^{1/2}$$

$$\frac{200^2}{609.6 \times 8.7} = 7.54$$

Hence

$$M = \left[1 + 0.6275\left(\frac{200^2}{609.6 \times 8.7}\right) - 0.003375\left(\frac{200^2}{609.6 \times 8.7}\right)^2\right]$$

$$M = 2.35$$

6.1.11 Example 3: assessment of axial dimensions – B31.G

The following example shows two areas of deep corrosion in a gas pipeline, detected by a magnetic inspection run and subsequently excavated for detailed measurement. The pipeline is made from X52-grade (358 MPa) 11.91 mm wall thickness pipe and has an outside diameter of 34 in. The current maximum operating pressure is set at 70 bar (70% SMYS). Assume these defects are located in a classification 1 area according to B31.8. Are these corrosion features acceptable to B31.G based on their axial dimensions (see Fig. 6.9)?

6.1.11.1 Corrosion defects

Step 1. Check that all required data are to hand: length, peak depth, material properties, operating pressure, diameter and wall thickness. Always remember to keep units consistent.

Step 2. Calculate the flow stress of the material, $\bar{\sigma} = 1.1 \times \text{SMYS}$, $\bar{\sigma} = 1.1 \times 358 = 393.8\,\text{N/mm}^2$.

Step 3. Is folias term $\sqrt{1 + 0.893\left(L_m/\sqrt{Dt}\right)^2} \leqslant 4.12$ in both cases:

$$M = \sqrt{1 + 0.893\left(\frac{250}{\sqrt{863.6 \times 11.91}}\right)^2} = 2.53\ (\textbf{defect 1})$$

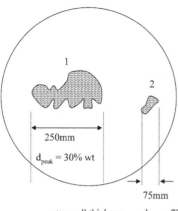

Figure 6.9 Reported corrosion features

$$M = \sqrt{1 + 0.893\left(\frac{75}{\sqrt{863.6 \times 11.91}}\right)^2} = 1.22 \ (\textbf{defect 2})$$

Step 4. Calculate the failure stress using the effective area equation

$$\frac{\sigma_f}{\overline{\sigma}} = \left[\frac{1 - (2d)/(3t)}{1 - (2d)/(3tM)}\right]$$

Therefore

$$\sigma_f = \overline{\sigma}\left[\frac{1 - 2/3 \times (3.57/11.91)}{1 - 2/3 \times (3.57/11.91) \times (1/2.53)}\right],$$

$$\sigma_f = 393.8 \times \frac{0.800}{0.921} = 342.06 \ \text{N/mm}^2 \ (\textbf{defect 1})$$

$$\sigma_f = \overline{\sigma}\left[\frac{1 - 2/3 \times (9.17/11.91)}{1 - 2/3 \times (9.17/11.91) \times (1/1.22)}\right],$$

$$\sigma_f = 393.8 \times \frac{0.486}{0.579} = 330.55 \ \text{N/mm}^2 \ (\textbf{defect 2})$$

It is also possible to calculate failure depths for different lengths of corrosion as

$$\frac{d}{t} = \frac{\bar{\sigma} - \sigma_f}{\frac{2}{3}[\bar{\sigma} - \sigma_f M^{-1}]}$$

Therefore, for defects 1 and 2 which have lengths of 250 mm and 75 mm respectively, if σ_f equals the operating stress with a safety margin of 1.39, then $PD/2t = (7 \times 863.6)/(2 \times 11.91) = 253.79 \times 1.39 = 352.77\,\text{N/mm}^2$.

The folias bulging factor M is calculated as

$$M = \sqrt{1 + 0.893 \left(\frac{250}{\sqrt{863.6 \times 11.91}}\right)^2} = 2.53 \text{ (\textbf{defect 1})}$$

$$M = \sqrt{1 + 0.893 \left(\frac{75}{\sqrt{863.6 \times 11.91}}\right)^2} = 1.22 \text{ (\textbf{defect 2})}$$

The calculated failure depths are

$$\frac{d}{t} = \frac{393.8 - 352.8}{2/3(393.8 - 352.8 \times 0.395)} = 0.24 \text{ (\textbf{defect 1})}$$

$$\frac{d}{t} = \frac{393.8 - 352.8}{2/3(393.8 - 352.8 \times 0.820)} = 0.59 \text{ (\textbf{defect 2})}$$

As shown above, the calculated failure depths for defects 1 and 2 are lower than the actual reported depths. This shows that neither failure depth is acceptable at a safety margin of 1.39 × MAOP. Consequently, the operator would have two choices:

- repair both defects;
- reduce the operating pressure.

It is important to note that this assessment assumes that the corrosion defects are not growing. In some instances where the corrosion defect is growing, the coating should be replaced at the site of the corrosion, and the cathodic protection levels should be checked to ensure that the pipe is fully protected.

6.1.12 Assessing the circumferential extent of corrosion

The previous corrosion example only considers loads created by hoop stress on the pipe, when in reality axial loads may also be present. Possible sources of axial load include:

- thermal expansion caused by high-temperature operation;
- ground movement;
- internal pressure.

In most cases the dominant stress on the pipe is the hoop stress, but consideration must be given to the circumferential extent of corrosion. A method of assessment was proposed by Kastner [20] for a part-through wall defect subject to an axial load (see Fig. 6.10).

In summary, the corrosion assessment approaches described should form part of an overall fitness-for-purpose assessment.

6.2 Assessing dents/profile distortions

Another common type of defect is profile distortion (dents and bulges). The significance of these features is that they increase the stress concentration and so reduce the burst strength of a pipe. The two main approaches when assessing the significance of dents are:

$$\sigma_l = \frac{PD}{4t}$$ $$\sigma_l = \frac{PD}{4t}$$

Figure 6.10 Circumferential corrosion assessment

- the static approach;
- the dynamic approach.

The static approach is a codified approach where the severity of a dent is based on its peak depth alone. The dynamic assessment approach utilizes the S–N approach to determine the remaining fatigue life of a dent using its peak depth. Before considering any of these approaches, it is important to define the different types of dent found on pipelines.

6.2.1 Plain dents

Plain dents contain no areas of increased stress concentration such as corrosion, gouge, weld or crack (see Fig. 6.11).

6.2.2 Smooth dents

These dents have a smooth profile and have no changes in wall thickness.

6.2.3 Kinked dents

Kinked dents show an abrupt change in curvature of the pipe wall, as shown in Fig. 6.12. These have similar shape characteristics to a buckle.

6.2.4 Dent–defect combination

Dent–defect combinations represent the most severe form of dents and contain other defects such as corrosion, gouges, cracking, etc. These result in lower failure pressures caused by:

- more global deformation;
- higher stress concentration;

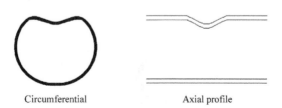

Circumferential   Axial profile

Figure 6.11   Plain dent

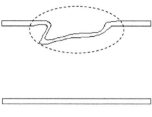

Axial Profile

Figure 6.12   Kinked dent

- possibility of cracks due to work hardening of the surface.

Dents can further be defined on the basis of whether they are free to re-round under the influence of internal pressure.

6.2.5 Constrained dents
Constrained dents are those that are not considered free to re-round under the influence of internal pressure. Typically, these dents are held in place by a rock (the indentor) and are subjected to fatigue loading. They are generally found at the bottom of the pipe, caused by construction during positioning of the pipe in the trench.

6.2.6 Unconstrained dents
These are free to re-round under the influence of internal pressure and are subjected to cyclic loading. Important characteristics of these dents are that:

- According to inspection data, they are usually found at the top of the pipe.
- As the indentor is removed, owing to the internal pressure, elastic springback occurs which reduces the depth of the dent.

6.2.7 Static assessment approach
Currently there are numerous codes that provide guidance on the maximum acceptable dent depth. However, there is slight

variation between the codes. The following list shows some of the codes that can be used to provide guidance on acceptable dent depth:

- API 1156 [21];
- EPRG [22];
- ASME B31.4 [7];
- ASME B31.8 [8].

It must be noted that the guidance given by most codes only covers plain dents. For dents associated with other features such as gouges, cracks or corrosion, most codes do not allow these combinations, and consequently there is little recommended guidance. Codes such as ASME B31.4 and B31.8 allow plain dents up to 6% diameter, whereas API 1156 suggests no limit on plain dents, as long as they are of the constrained type.

Dents associated with welds can result in low burst pressures, and are susceptible to cracking at the weld toe. Industry operators have conducted a number of tests on dents associated with welds and reported some specimens exhibiting very low failure stresses [23]. API 1156 [21] suggests that dents on welds of up to 2% pipe diameter are acceptable provided the weld material is of moderate to high toughness. Some pipeline operators follow this guidance, but ultimately the dent must be checked for the presence of cracks using non-destructive testing.

In any fitness-for-purpose assessment it is important to determine the location of the reported dent since this can give an indication of the damage mechanism, i.e. is it located at the bottom of the pipe or at the top? Dents located at the top of the pipe are frequently caused by third party damage and may contain other defects. In addition, these are likely to be unconstrained, so they should be assessed for fatigue.

6.2.8 Fatigue life estimation of dents
Dents are assessed for fatigue because they create a disturbance in the curvature of the pipe wall, resulting in

an area of increased stress concentration. Under fatigue loading conditions, the depth of the dent may change (see Fig. 6.13). The important parameters when assessing dents are as follows:

- Because of re-rounding effects, the depth of a dent created at zero internal pressure will be different under pressure loading conditions.
- Dent depth changes under different internal pressure conditions (i.e. pressure cycling).

This effect is known as re-rounding, which changes under different fatigue loading conditions. As shown in Fig. 6.13, as the pressure increases, the dent depth becomes smaller, whereas under lower pressure conditions the dent depth is larger. This causes fluctuating stresses around the dented region. If fatigue loading is high enough, the dent will ultimately fail.

A common method for estimating the fatigue life of steels is through the basic S–N curve, displaying stress range versus number of cycles. This basic approach has been utilized within the pipeline industry, applying the basic S–N curve and relating this to the increased stress concentration due to the dent.

A number of models have been developed using a semi-empirical method for predicting the fatigue life of plain dents. These are based on the use of an expression for stress concentration due to the dent. The fatigue life is then calculated using the basic S–N curve. There has been

Figure 6.13   Re-rounding effects on dents

extensive fatigue testing of dents conducted by a number of organizations, including:

- Battelle [24];
- CANMET [25, 26];
- British Gas [27, 28];
- EPRG [29, 30];
- SES [31, 32];
- API 1156 [22].

Alongside these tests, a number of elastic–plastic finite element analyses were conducted. Empirical and semi-empirical methods to assess the fatigue strength of plain dents have been developed by the European Pipeline Research Group (EPRG). These methods are based on a basic design S–N curve for plain pipe material, with fatigue life being calculated by taking into account the stress concentration due to the dent.

Basic fatigue design S–N curves include:

- PD 5500 [33];
- DIN 2413 [34].

EPRG/PRCI [29, 30] have developed an empirical method for predicting the fatigue life of a dent. Fatigue life is calculated using the DIN 2413 fatigue curves modified using a relevant stress concentration factor. The fatigue life of a plain dent is then calculated using the following expression:

$$N_C = 100[(\text{UTS} - 50)/(2\sigma_A K_S)]^{4.292}$$

where $2\sigma_A = \sigma_U \left[B\left(4 + B^2\right)^{0.5} - B^2 \right]$, the equivalent cyclic stress at $R = 0$, and

$$B = \frac{\sigma_a/\sigma_U}{\{1 - \sigma_a/\sigma_U[(1 + R)/(1 - R)]\}^{0.5}}$$

where R is the ratio of the minimum stress/maximum stress in the fatigue cycle, σ_a is the cyclic stress (N/mm^2) and σ_U is the ultimate tensile strength (N/mm^2). Here

$$R = \frac{\sigma_{min}}{\sigma_{max}}$$

$$K_S = 2.871\sqrt{K_d}$$

Hence, the stress concentration due to the dent is

$$K_d = H_0 \frac{t}{D}$$

Taking into account the re-rounded depth due to the internal pressure:

$$H_0 = 1.43 H_r$$

The citation to this reference conveys no rights to the reader in the material referenced and it may not be used without the prior written permission of Pipeline Research Council International, Inc.

Currently, the best technologies for detecting dents are high-resolution geometric tools that give an accurate estimate of peak depth. Unfortunately, these tools alone do not provide all the required information for an accurate dent assessment. Wall thickness, orientation and relative distance from a weld are also important parameters. Hence, a geometric inspection tool usually follows intelligent tools such as:

- a magnetic inspection tool;
- an ultrasonic inspection tool;
- a gauge pig.

Consider the following example of a static dent assessment.

A natural gas pipeline was inspected using ultrasonic and geometric inspection tools, and showed the dent illustrated in Fig. 6.14. Using the guidance provided on static dent assessments for pipelines, what recommendations would you make?

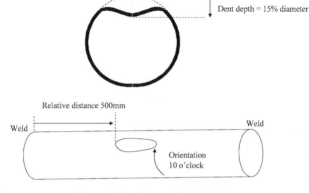

Figure 6.14 Reported dent feature

Step 1. Is the dent plain? Yes, this is not associated with a seam weld, and is therefore a smooth plain dent. As discussed earlier, dents associated with welds can have low burst pressures and are generally recommended for repair.

Step 2. What is the peak depth of the reported dent? In this case the maximum depth is 15% of diameter. Most codes would not allow a dent of this size and would recommend a repair. In addition, this dent is located in the top half of the pipeline, so two conclusions can be made:

- The dent may be unrestrained.
- The dent may also have been caused by third party damage.

Since B31.8 [8] recommends that dents up to 6% would be acceptable, this dent would be recommended for repair. However, a further check on this dent would be to calculate the remaining fatigue life.

Step 3. Codes such as API 1156 [21] recommend that unrestrained dents greater than 2% should be assessed for fatigue. Hence if the pipeline had numerous reported dents, then as previously discussed, dents at the top and bottom of the pipeline can be separated for assessment. In summary, API 1156 [21] suggests that:

- Dents at the top of the pipe are likely to be unrestrained and subjected to fatigue loads.
- Dents at the bottom of the pipe are likely to be constrained (i.e. held in place by the indentor) and not subjected to fatigue loads.

In conclusion, based on a static assessment using most codes, this type of dent would be unacceptable.

6.3 Significance of manufacturing and construction defects

The main types of manufacturing defect include:

- wall thickness variations;
- indentation damage;
- laminations;
- inclusions.

The main types of construction defect include:

- girth weld defects (lack of fill or misalignment);
- weld cracks;
- porosity;
- denting.

As this shows, there are numerous types of defect to consider, but some of the more serious types of defect are laminations and cracks (i.e. planar-type defects). Lamination features can bulge when operating in sour conditions and also start to form hydrogen-induced cracking (see Section 5.4.6). In addition, planar-type defects such as cracks can grow under fatigue conditions. Code methods are used to assess planar defects.

6.3.1 Assessing cracks in pipeline steel

Many older pipeline steels have low toughness material properties and can be susceptible to cracks forming under fatigue conditions. Most of the new pipeline steels, however, are now made with higher-toughness steels and are able to withstand higher critical defect sizes. Areas susceptible to cracks are the seam weld or girth weld, particularly in low-toughness weld material. Currently, the best ways to detect cracks in a pipeline are by intelligent inspection using ultrasonic or special crack detection tools, or alternatively excavation and non-destructive testing using ultrasonics or magnetic particle inspection. Table 6.4 shows the main technologies used throughout the pipeline industry to detect crack-like defects.

At present, the main codes used for assessing cracks are:

- BS 7910 [**35**];
- API 579 [**36**].

Table 6.4 Main inspection methods for detecting crack-type defects

Inspection method	Crack detection capability
Magnetic inspection	Primarily used for detecting corrosion, but can detect girth weld cracking
Ultrasonic inspection	Crack detection tools can be used for detecting cracks and laminations in both oil and gas pipelines
Transverse field inspection	This is capable of detecting seam weld defects. It uses the same principle as MFL inspection but magnetizes the pipe so that axial anomalies such as internal channelling corrosion or seam weld anomalies can be detected
Eddy current	A magnetic field is induced through the pipe wall, creating an eddy current. This type of tool is capable of detecting axial cracking and SCC

Both these documents are used to assess flaws in welded structures using a failure assessment diagram (FAD) to determine the acceptability of cracks. There are three levels of assessment, 'level 1', 'level 2' and 'level 3', each increasing in complexity and based on different amounts of input data. While both API 579 and BS 7910 use three levels of assessment, they are not identical.

Within BS 7910 [35] and API 579 [36] the FAD is used as a method of taking into account the applied stress, geometry and fracture toughness of the pipeline. This type of diagram is used in levels 1 to 3 of BS 7910 to determine the acceptability of cracks by plotting a point on the diagram (see Fig. 6.15). A crack-type defect is considered acceptable or unacceptable depending upon whether this lies within an acceptable boundary.

The assessment level used depends on the input data

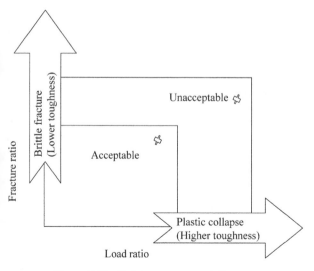

Figure 6.15 Failure assessment diagram

available and the conservatism required. The levels can be summarized as follows:

- Level 1 is a simplified assessment method when there are limited data on material properties.
- Level 2 is the normal assessment route.
- Level 3 is based on complex ductile tearing resistance analysis.

As shown in Fig. 6.15, crack-type defects have two failure mechanisms:

- plastic collapse;
- brittle fracture.

When the material approaches higher toughness values, it is predicted that a defect will fail by plastic collapse. However, when the material approaches lower toughness values, it is predicted that a defect will fail by brittle fracture. In addition to material properties, it is important to have accurate defect dimensions. Figure 6.16 shows basic terms used for defect measurements, where $2c$ is the defect length, a is the defect depth and B is the section thickness. In order to determine a point on this diagram, values for load ratio and fracture ratio must be calculated.

6.3.2 BS 7910 level 1 simplified assessment

When a conservative estimate is required and there is limited information on material properties, the BS 7910 level 1 assessment method is often used. In addition, this simplifies

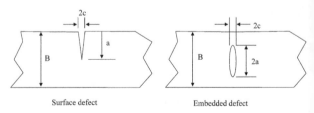

Surface defect   Embedded defect

Figure 6.16 Basic terms for crack assessment

the material stress–strain curve by approximating it to being perfectly elastic–plastic with no strain hardening effect. Only in levels 2 and 3 is stress–strain data required for the actual material curve. When determining a point on the failure assessment diagram, values for both K_r and L_r, the fracture and load ratio respectively, are required. These are calculated using the following equations:

$$\frac{K_I}{K_{mat}} = K_r$$

$$\frac{\sigma_{ref}}{\sigma_f} = L_r$$

As shown in Fig. 6.17 (level 1 FAD), the flaw would be acceptable if K_r were less than 0.707 and L_r were less than 0.8.

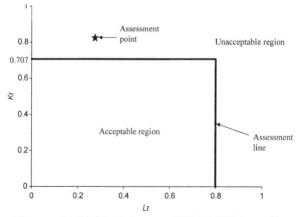

Figure 6.17   Level 1 failure assessment diagram

How is the material fracture ratio K_r calculated? This equation is based on the ratio of the stress intensity factor to the material fracture toughness, and is calculated using the following equation (see BS 7910 [**35**]):

$$K_I = (Y\sigma)\sqrt{(\pi a)}$$

where $Y\sigma$ is dependent on the geometry of the defect, the maximum stress and whether the defect is surface breaking or embedded:

$$Y\sigma = Mf_w M_m \sigma_{max}$$

M is the bulging correction factor, f_w is the correction term in the stress intensity factor for elliptical flaws (which varies depending on whether the flaw is an internal or embedded defect), M_m is the stress intensity magnification factor (calculated on the basis of geometric dimensions a and c), σ_{max} is the maximum tensile stress [calculated using $\sigma_{max} = k_t S_{nom} + (k_m - 1)S_{nom} + Q$, k_t is the stress concentration factor, k_m is the stress concentration due to misalignment and Q = any secondary stresses.

In order to determine the material fracture toughness K_{mat}, firstly it has to be decided whether fracture toughness data are available. If there are no direct measurements of fracture toughness, the approach described in BS 7910 [**35**] details calculation of a toughness value based on the Charpy V-notch energy. This provides a conservative lower-bound correlation applicable to a wide range of steels:

$$K_{mat} = \frac{820\sqrt{C_v} - 1420}{B^{1/4}} + 630$$

where K_{mat} is the the lower-bound material fracture toughness (N/mm$^{3/2}$), C_v is the Charpy V-notch energy at the operating temperature (J) and B is the the section thickness for which K_{mat} is required (mm).

How is the load ratio L_r calculated? The load ratio is based on the reference stress σ_{ref} and the flow stress σ_f, and is calculated as follows:

$$\frac{\sigma_{ref}}{\sigma_f} = L_r$$

For a level 1 assessment, flow stress is taken as an average of the yield and ultimate tensile strength of the material, $(\sigma_Y + \sigma_U)/2$, up to a maximum of $1.2\sigma_Y$. The reference stress σ_{ref} is dependent on whether the defect is located internally or externally and is calculated using the appropriate stress solutions in BS 7910 [35] for bending and membrane stresses.

6.3.3 BS 7910 level 2 normal assessment method

There are two slightly different approaches within the level 2 assessment method (level 2A and level 2B). Both use the FAD, but level 2B requires significantly more data in the form of a specific stress–strain curve (using true stress–strain). Level 2A utilizes a generalized FAD not requiring actual stress–strain data. For this approach, the FAD is represented by an equation of the curve, as shown in Fig. 6.18.

As shown in Fig. 6.18, cut-off is provided to prevent localized plastic collapse where $L_{r\,max} = (\sigma_Y + \sigma_U)/(2\sigma_Y)$. Similar to the level 1 assessment, in order to find a point on the FAD, values for K_r and L_r are calculated. If the point lies within the area of the line, the defect is considered acceptable. If it lies beyond the line, the defect is considered unacceptable. The main difference with this approach lies in the calculation of K_r and L_r.

The stress intensity factor K_I is calculated as before, where $Y\sigma$ is dependent on the geometry of the defect. The level 2A assessment defines $Y\sigma$ as

$$Y\sigma = (Y\sigma)_P + (Y\sigma)_S$$

where $(Y\sigma)_P$ and $(Y\sigma)_S$ are the primary and secondary stress contributions respectively. These are calculated as

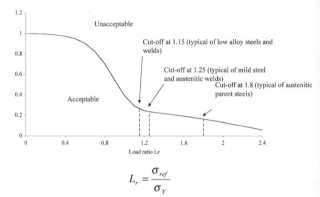

$$L_r = \frac{\sigma_{ref}}{\sigma_Y}$$

Figure 6.18   Level 2 failure assessment diagram

follows:

$$(Y\sigma)_P = Mf_w[k_{tm}M_{km}M_mP_m + k_{tb}M_{kb}M_b[P_b + (k_m - 1)P_m]$$

and

$$(Y\sigma)_S = M_mQ_m + M_bQ_b$$

where k_{tm} is the membrane stress concentration factor due to major discontinuity, M_{km}, M_m, M_b and M_{kb} are the stress intensity magnification factors, P_m and Q_m are the primary and secondary membrane stresses P_b and Q_b are the primary and secondary bending stress, k_{tb} is the bending stress concentration factor and k_m is the stress concentration due to misalignment.

6.3.4 API 579 assessment levels

Like BS 7910, the fitness-for-service code API 579 [36] can be used to assess crack-like flaw geometries for either surface breaking or embedded defects. A brief description of each level is as follows.

The level 1 assessment is used for assessing cracks that are away from other structural discontinuities, and is based on a simple screening assessment. It should be used if the only loading on the pipe is due to internal pressure and there are no bending loads. Figure 6.19 shows the chart used to conduct a simple screening assessment to find the maximum acceptable crack length for a through-wall defect and a quarter-wall defect.

As shown in Fig. 6.19, this assessment can be done for a defect located in the base metal, or in weld material that has been subjected to post-weld heat treatment (PWHT). The axes of this diagram show crack length, $2c$, against material toughness represented by a reference temperature and the operating temperature.

The API 579 level 2 assessment is very similar to the level 2

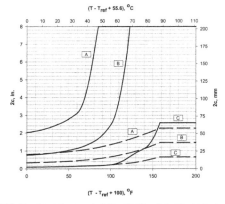

Definition of screening curves (solid line 1/4-t flaw, dashed line 1-t flaw):
A – Allowable flaw size in base metal.
B – Allowable flaw size in weld metal that has been subject to PWHT
C – Allowable flaw size in weld metal that has not been subjected to PWHT
PWHT - Pre-weld heat treatment

Figure 6.19   API level 1 assessment diagram

normal assessment in BS 7910, which requires the calculation of the load ratio L_r and the fracture ratio K_r, to determine a point on the FAD. In addition, partial safety factors can be used for variables such as flaw size, material fracture toughness and stress. This is primarily used to assess flaws located at a structural discontinuity. Furthermore, detailed material properties are required and a determination of the current stress state such as by numerical analysis, or using code equations. Consequently, API 579 levels 2 and 3 assessments only apply if there are additional loading conditions (other than the membrane stress), such as bending loads.

An API 579 level 3 assessment should be used if a flaw is expected to grow during service. This method also uses the FAD, but there are five approaches to choose from:

- the use of partial safety factors based on risk or a probabilistic analysis;
- the construction of the FAD diagram using specific stress–strain material data;
- the use of the FAD diagram alongside actual loading conditions;
- the use of a ductile tearing analysis where the fracture tearing resistance is defined as a function of the amount of stable ductile tearing;
- the use of other recommended assessment codes (see API 579, Section 1 [**36**]).

6.3.5 Laminations in pipeline steel

Defects introduced in the pipeline during manufacturing include wall thickness variations, laminations and metal loss due to skimming contact of the surface. Lamination features should have been detected during the pipe manufacturing process, but can often still be found following a pipeline inspection (see Fig. 6.20). If operating in sour conditions, it is possible that laminations can grow or bulge during service, so they must be continuously monitored using repeat internal inspections to identify any significant changes in their size.

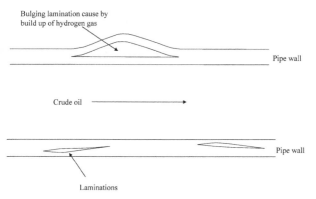

Figure 6.20   Lamination features

When determining the immediate and future integrity of laminations, account must be taken of the following:

- Is the pipeline operating in sour conditions?
- Are the laminations sloping or parallel to the pipe surface?
- Are the laminations mid-wall or surface breaking?
- Are they associated with any other feature such as a girth weld?

The API 5L specification for line pipe [9] sets a limit on laminations found following manufacture. Any lamination with a minor dimension exceeding 19 mm and an area greater than 7742 mm^2 is deemed unacceptable. Methods are available to assess these features, such as API 579 [36]. The assessment method within API 579 is again divided into different levels of assessment:

- Levels 1 and 2 apply if the laminations are parallel to the surface and have no through-thickness cracking.
- A level 2 approach is required if the lamination is operating in a hydrogen environment (i.e. if the product contains hydrogen sulphide).
- A level 3 approach is required if the lamination is both

located at a structural discontinuity such as a weld and is also operating in a hydrogen environment.

- Laminations that are not parallel to the pipe surface should be assessed as a crack-like flaw as described in Section 6.3.1.

In summary:

Level 1

A lamination is considered acceptable if all the following are satisfied (see Fig. 6.21);

- The pipeline is not operating in a hydrogen environment.
- The distance from the edge of the lamination to the nearest major structural discontinuity is $>1.8\sqrt{Dt_{nom}}$.
- If these conditions are not met, then a level 2 approach is required.

Level 2

If operating in a hydrogen environment, a lamination would be considered acceptable if the following are satisfied:

- The distance from the edge of the lamination to the nearest major structural discontinuity is $>1.8\sqrt{Dt_{nom}}$.

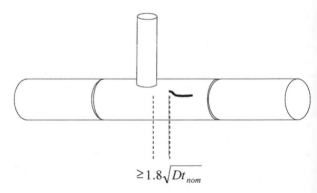

$$\geq 1.8\sqrt{Dt_{nom}}$$

Figure 6.21   level 1 requirements

- Check for the presence of cracks. If the distance between the lamination and the nearest weld is <25.4 mm or twice the wall thickness, then it is considered to be associated with the weld. In addition, there is a risk of cracks developing that may propagate along the weld line in the through-thickness direction. If these cracks are found, a level 3 assessment is required (or a permanent repair).

Level 3

A level 3 analysis is only required if a reported lamination is oriented in the through-thickness direction or where cracks are reported to have developed at the weld. This involves conducting a detailed stress analysis using numerical analysis, or fracture mechanics methods.

Reproduced courtesy of the American Petroleum Institute and may not be used without prior written permission.

By definition, a lamination is a type of manufacturing feature. However, as previously discussed, other types of manufacturing defect include wall thickness variations and inclusions.

6.3.6 Manufacturing defects

Wall thickness variations are commonly found in seamless pipe, especially in the older designs. Specifications such as API 5L for seamless pipe [9] allow a maximum of 12.5% reduction on wall thickness. With wall thickness variation, the operator must consider whether these variations are acceptable (see Fig. 6.22).

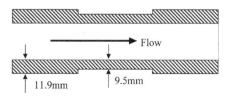

Flow

11.9mm   9.5mm

Figure 6.22   Wall thickness variations

Inclusions are a discontinuity (e.g. an area of manganese sulphide) in the microstructure of a steel. They can cause cracking if atomic hydrogen diffuses into these areas when operating in sour operating conditions, e.g. hydrogen-induced cracking (see Section 5.4.6). When operating in these conditions, a pipeline should be monitored through repeat inspections (see Fig. 6.23).

6.3.7 Fatigue assessment
Pipelines are often subjected to cyclic pressure variations. These variations can cause defects to fail at stress levels lower than the yield strength for a static load and are caused by:

- seasonal changes in demand;
- batching operations;
- daily pressure fluctuations;
- valve operations (opening and closing).

When conducting fatigue calculations, there are two main approaches: the S–N approach and the fracture mechanics approach (see Fig. 6.24). When a pipeline is subjected to fatigue loading, defects that should be considered for a fatigue assessment include:

- seam weld defects;
- girth weld defects;
- cracks and laminations;
- dents;

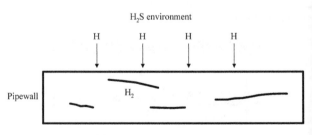

Figure 6.23 Inclusions in pipeline steel

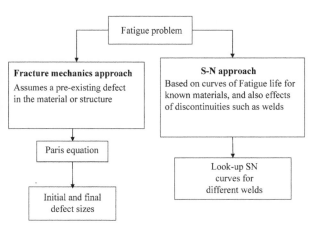

Figure 6.24   Fatigue assessment approaches

- gouges;
- other (patch repairs).

First, the *S–N* approach. If steel specimens are fatigue loaded at a number of different stress levels until failure occurs, and a plot of applied stress to logarithm number of cycles for failure is produced, this results in what is known as the *S–N* curve. A characteristic of these curves is that there is a limiting stress level below which fatigue failure will not occur. This is known as the fatigue limit or endurance limit and is usually between 35 and 60% of the tensile strength (see Fig. 6.25).

Note that *S–N* curves are based on a best fit of test data. Scatter is often present in these data owing to uncertainties such as material properties, loading conditions, test preparation, etc. Areas of increased stress concentration such as welds, holes or other defects can significantly reduce the fatigue life (see Fig. 6.26). This *S–N* approach is generally used at the design stage for pressure vessels and is described in codes such as PD 5500 [**37**]. This code provides fatigue

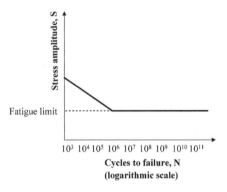

Figure 6.25   S–N fatigue diagram

curves for different classes of welded joints. To conduct a defect assessment on features (i.e. a pre-existing defect) such as cracks and laminations, a fracture mechanics approach is required.

Failure due to fatigue consists of three identifiable phases:

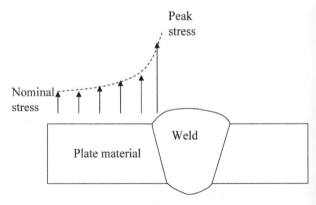

Figure 6.26   Stress profile associated with a weld

- crack initiation (phase 1);
- crack propagation/growth (phase 2);
- failure (phase 3).

Cracks formed by fatigue usually initiate at points of increased stress concentration such as welds, dents, manufacturing features or other forms of damage. Experimentally it is possible to measure crack length during cyclic stress and plot this as crack length, a, versus number of cycles, N (see Fig. 6.27).

The message of Fig. 6.27 is that crack growth rate can be calculated at any instant by taking the slope da/dN. Note that the crack growth rate is initially small but increases significantly with increasing number of cycles, leading to failure at a_f. Since crack growth rate is a function of applied stress, crack size and material properties, da/dN can be represented as a function of the stress intensity factor ΔK. Consequently, the remaining fatigue life of a crack can be calculated using fracture mechanics. These terms are included in what is known as the Paris law equation [36].

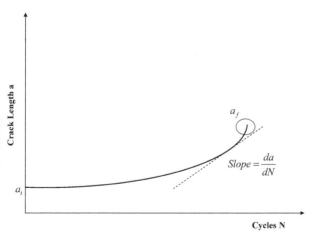

Figure 6.27   Crack length versus number of cycles

This equation is published in codes such as Section 8 of BS 7910 [35]:

$$\frac{da}{dN} = A(\Delta K)^m$$

where the stress intensity factor range at the crack tip is $\Delta K = Y(\Delta\sigma)\sqrt{\pi a}$, $\Delta\sigma$ is the cyclic stress, a is the instantaneous crack depth, Y is dependent on crack geometry and m and A are constants for a particular material and based on the applied conditions. Using BS 7910 [36], these values are $m = 3$ and $A = 5.21 \times 10^{-13}$ and assume the following:

- yield strength $< 600\,\text{N/mm}^2$;
- Operating temperature $< 100\,^\circ\text{C}$.

Figure 6.28 shows that if a plot of log stress intensity factor versus log crack growth rate is produced, this shows a linear portion in which the crack growth rate da/dN can be predicted [35].

Since $da/dN = A(\Delta K)^m$, it is possible to develop an expression where the number of cycles to grow from an initial to final defect size can be calculated. This approach is useful for predicting the fatigue life of defects including cracks and laminations. The expression is derived in the folllowing way.

Since $\Delta K = Y(\Delta\sigma)(\pi a)^{0.5}$, it follows that

$$\frac{da}{dN} = A\left(Y\Delta\sigma\sqrt{\pi a}\right)^m$$

If this equation is rearranged in terms of defect size and fatigue life, and then integrated between the initial and final defect sizes it is expressed as:

$$\frac{da}{dN} = A\,Y^m\Delta\sigma^m\pi^{0.5m}a^{0.5m}$$

$$\int_{a_i}^{a_f} a^{-0.5m}\,da = A\,Y^m\,\Delta\sigma^m\,\pi^{0.5m}\int_0^N dN$$

$$\left[\frac{a^{(1-0.5m)}}{1-0.5m}\right]_{a_i}^{a_f} = A\,Y^m\,\Delta\sigma^m\,\pi^{0.5m}\,N$$

$$\left[\frac{a_f^{(1-0.5m)}}{1-0.5m}\right] - \left[\frac{a_i^{(1-0.5m)}}{1-0.5m}\right] = A\,Y^m\,\Delta\sigma^m\,\pi^{0.5m}\,N$$

$$\left(a_f^{(1-0.5m)} - a_i^{(1-0.5m)}\right)\left(\frac{1}{1-0.5m}\right) = A\,Y^m\,\Delta\sigma^m\,\pi^{0.5m}\,N$$

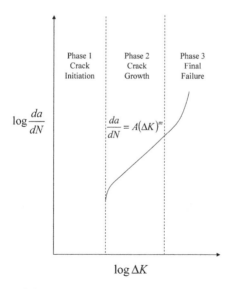

Figure 6.28 Stress intensity factor versus log crack growth rate

Rearranging in terms of fatigue life N yields

$$N = \frac{a_f^{(1-0.5m)} - a_i^{(1-0.5m)}[1/(1-0.5m)]}{A Y^m \Delta \sigma^m \pi^{0.5m}}$$

Therefore

$$N = \frac{a_f^{(1-0.5m)} - a_i^{(1-0.5m)}}{(1-0.5m)[A(Y\Delta\sigma)^m \pi^{0.5m}]}$$

This equation is useful in calculating the fatigue life of cracks growing from an intial size a_i to a final defect size a_f (see Fig. 6.29).

Another type of defect that also has the possibility of producing cracks are gouge-type defects. These can seriously affect the burst strength of a pipeline and are often caused by mechanical damage or third party damage.

6.3.8 Gouges

A gouge is by nature an area of metal loss. It is, however, treated differently to corrosion as, by its nature, a gouge

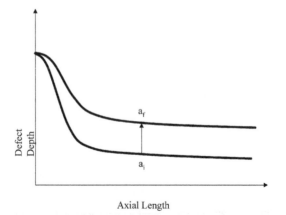

Figure 6.29 Crack growth from initial to final defect size

creates a work-hardened layer on the surface, which reduces the ductility of the material.

From inspection data alone, it is not usually possible to see whether a gouge has cracking associated with it. Gouges should therefore be treated with caution and should be excavated and examined to check for the presence of cracks. This can often be done through non-destructive methods such as magnetic particle inspection. There is various guidance in oil and gas codes on what to do with gouges, but it can be very general and will not always tell you whether they are acceptable in terms of immediate and future pipeline integrity.

Solutions to gouges that are used throughout the pipeline industry include:

- dressing of the gouge to remove the work-hardened layer;
- permanent repair;
- assessment as an area of metal loss.

Finally, in order to conduct a detailed assessment of defects using the various approaches, it is important to have accurate pressure data, and to understand the maximum and minimum pressure variations. Pipelines are often subjected to fatigue loads, so analysis and interpretation of cyclic pressure data is essential.

6.3.9 Pressure cycling data

Ideally, fatigue data should be recorded on an hourly basis during both winter and summer operations. Figure 6.30 shows an example of a cyclic pressure spectrum. As can be seen, this is quite complex and can be confusing. A method of simplifying this is to convert it into blocks of constant amplitude. This can be achieved by using cycle counting methods that simplify the pressure spectrum data. One such approach is the 'rainflow' counting approach.

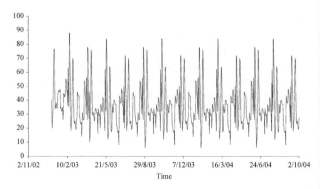

Figure 6.30   Cyclic pressure data

6.3.10 Rainflow counting

As implied by the name, the pressure spectrum is imagined as a set of peaks and troughs over which rain falls. There are two basic rules that need to be followed (see Fig. 6.31).

The figure shows a number of cycles based on the paths A to J. Note that each path is equivalent to one half-cycle (i.e. path C–D is one half-cycle). There are seven half-cycles which are determined as follows:

- Path A–B is formed from rain flowing from trough A and stopping at B where it reaches a trough more negative than its starting point.
- Path C–D starts at peak C and continues flowing to D since no other peak more positive than its starting point is reached.
- Path E–F starts at trough E but encounters a trough more negative than its starting point, so it stops at F.
- Path G–H starts at peak G and stops at H when it encounters rain falling vertically.
- Path I–J starts at trough I and continues flowing to J as it does not reach a trough any more negative than its starting point.

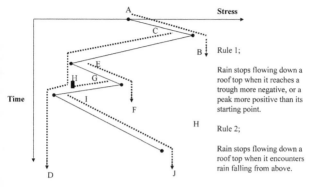

Figure 6.31   Rainflow counting method

Each path represents a half-cycle at a particular stress range. Using the Minor's rule, we can now reduce the number of cycles in each stress range into an equivalent number of cycles at the maximum stress range. The maximum stress range is therefore chosen, since this would obviously be the most conservative approach. This is done using the following expression:

$$n_1 s_1^m = n_2 s_2^m$$

where n is the number of cycles and m is the slope of the design S–N curves (typically 3.0 for steel).

Chapter 7

Pipeline Condition Monitoring and Repair Methods

Above-ground pipeline monitoring techniques are an essential part of maintenance activities. The primary objectives are to:

- highlight problem areas where the levels of cathodic protection are ineffective;
- find areas of coating damage.

Figure 7.1 shows the main above-ground survey methods.

Inspection of offshore pipelines is more complicated. In this case the following methods are often used:

- side-scan survey;
- remote operated vehicles (ROVs);
- visual inspection using divers;
- intelligent inspection tools.

7.1 Pearson survey

This method is used mainly for detecting coating defects and requires two operators (see Fig. 7.2). As shown, a signal loop is created, as both operators are electrically connected through the following items:

Figure 7.1 Above-ground survey methods

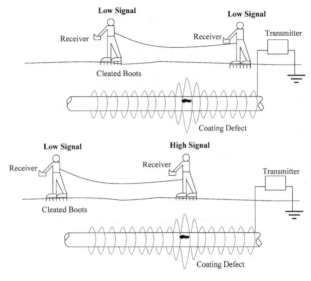

Figure 7.2 Pearson survey

- transmitter;
- pipeline;
- soil;
- cleated boots (boots with steel spikes attached underneath);
- a connecting wire between the operators.

The basic method of operation is that an alternating current (a.c.) signal is passed through the pipeline. This a.c. signal is then detected using audio headphones and a measurement read-out. At the area of open coating defects the signal increases, detected as each operator passes over the defect. Limitations of this approach are that:

- It cannot detect disbonded coating (i.e. is only effective at open coating defects).
- It does not indicate whether the CP levels are adequate.

- It cannot be used accurately to size a coating defect.
- It is usually employed on short sections of pipeline.

7.2 CIPPS survey

A close interval polarised potential survey (CIPPS) is conducted along the entire length of the pipeline and is often used in conjunction with other coating defect surveys. The technique is able to detect coating defects, but it is primarily used to measure the effectiveness of the cathodic protection (CP) system. It measures two important parameters:

- CP 'on' potential levels;
- CP 'off' potential levels.

These 'on' and 'off' potential measurements are made during synchronous interruption, which alternates between these two measurements. What is the difference between these measurements? As discussed in Section 3.3, potential measurements are always taken with reference to an electrode buried in the soil. It is important to measure the potential at the surface of the pipe (i.e. the polarized potential), hence 'off' potential measurements are made which take into account the soil resistance.

During a survey, readings are taken at close intervals of ~2 m. As shown in Fig. 7.3, a single operator is connected to a test post via two training wires. The test post is also electrically connected to the pipe. As the operator travels along the pipeline route, measurements are taken above the pipe level, using an electrode that the operator holds. Figure 7.4 shows the typical results of a CIPPS survey. NACE recommended practice [11] states that for an effective CP system the target 'off' potential level along the pipeline should be maintained within a potential range of −850 mV to −1200 mV.

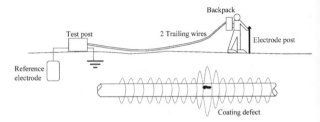

Figure 7.3   CIPPS survey

7.3 DCVG survey

The direct current voltage gradient (DCVG) survey technique was developed and used in Australia and, like the Pearson survey, primarily locates coating defects. This method differs, however, in that it provides accurate results for sizing of coating defects. It works by using an aboveground operator to measure changes in voltage along the pipe (i.e. the voltage gradient). These voltage gradients occur at the site of coating defects, and these DCVG measurements are based on percentage internal resistance (%IR). Hence, the severity of coating defects is categorized on the basis of the internal resistance drop:

- category 1: 1–15%IR (least severe);
- category 2: 16–35%IR;
- category 3: 36–60%IR;
- category 4: 61–100%IR (worst case).

By combining survey techniques, it is possible to find areas where:

- Significant coating defects exist.
- CP levels are inadequate.
- Both CP levels are low and coating defects exist, i.e. areas of potential corrosion.

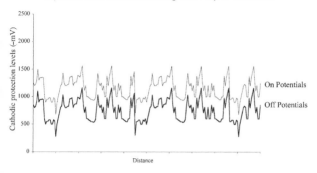

Figure 7.4   Typical layout of CIPPS survey results

7.4 Repair techniques

A pipeline repair must provide structural support and, where applicable, contain the internal pressure. Currently there are numerous types of repair used throughout the pipeline industry, but they can be broadly categorized into two main types:

- permanent repairs;
- temporary repairs.

A summary of the different repair methods is as follows:

- grinding repair;
- cut-out and replacement;
- temporary/leak clamp repair;
- epoxy sleeve repair;
- stopple and bypass operation;
- snug-fitting sleeve repair;
- stand-off sleeve repair;
- patch repair and weld deposition;
- composite sleeve repair;
- mechanical clamps;
- offshore stopple and bypass operation.

7.4.1 Grinding repair

Codes such as ASME B31.4 [7] and ASME B31.8 [8] allow the use of grinding to smooth the defect profile. This repair method is often used to smooth out areas of increased stress concentration. As discussed in Section 6.3.8, gouges contain a work-hardened layer at the surface, so grinding is sometimes used to remove this layer and eliminate points of increased stress.

7.4.2 Cut-out and replacement

This approach is recommended in codes such as ASME B31.8 [8]. It is generally used as a last resort since it causes the most disruption and cost for the pipeline operator as the pipeline must be depressurized, decommissioned and the product appropriately disposed of. This can be very expensive and time consuming.

7.4.3 Temporary/leak clamp repair

This consists of two shell halves that are bolted together over the damaged section of pipeline. It is generally used as a temporary repair for leaking defects, and is usually replaced within a year following application.

7.4.4 Epoxy sleeve repair

Epoxy sleeve repairs consist of two half-shells positioned around the diameter of the pipeline. Both ends of the sleeve are sealed using a putty mixture, and the shell is filled with epoxy. The epoxy is injected into the repair under pressure, which forces the mixture completely to encase the inside of the shell. Usually the mixture is filled from the bottom of the shell, which allows any trapped air to escape. Once the epoxy mixture has cooled, it forms an extremely high-stiffness material that prevents any further deformation of the pipe.

A major benefit of this repair method is that there is no need for direct welding onto the pipe. Preparation of the repair and damaged region is as follows:

- The surface of the pipeline and each shell half are shot blasted to ensure a clean surface. This has two purposes:

firstly to ensure that the surface is free from grit and dirt, and secondly the shot blasting process creates a rough surface texture, increasing the surface area and improving the bonding with the epoxy.

- The two halves of the repair are placed around the circumference of the pipe ready for setting the annular gap. This gap is adjusted to ensure that the shells are seated evenly around the pipe and that no bending stresses occur as a result of misalignment.
- The two halves are welded together and checked for any defects. This is the most important part of the process, since a defective weld could fail during operation.
- A putty mixture is applied to the ends of the shell. This hardens and creates a seal at the ends of the repair.
- Finally, using specialist pumping equipment, the epoxy mixture is injected into the shell. Note that different grades of epoxy are used for applications during winter and summer seasons.

7.4.5 Stopple and bypass operation

When major leaks occur that cannot be repaired using a temporary clamp, the stopple bypass operation is often used. This provides a means of bypassing the flow around the defective area of pipe (see Fig. 7.5) and involves directly welding a full encirclement tee onto the pipeline. Initially, the pipeline is drilled using a boring device which drills a hole through the pipe wall and removes the remaining steel coupon.

7.4.6 Snug-fitting sleeve repair

The main purpose of this repair is snugly to fit a repair section around the pipe and so prevent bulging. Fillet welds are applied by direct welding onto the pipe, sealing each shell half to the pipeline (see Fig. 7.6).

This repair method provides structural support and is used to repair defects located in the main pipe body, girth welds and seam welds.

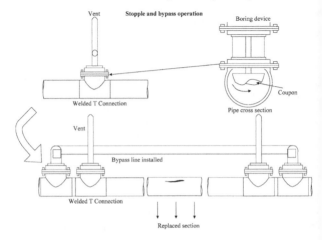

Figure 7.5 Stopple and bypass operation

7.4.7 Stand-off sleeve repair

This repair method is used for repairing curved or distorted sections of pipe. It is similar to the snug-fit type shell, but is connected to the pipeline using a welded split collar at each end of the stand-off shell. It is important that the material strength used for both these repair methods is at least equivalent to the grade of steel used for the pipeline.

7.4.8 Patch repair and weld deposition

Patches and weld depositions are used for repairing non-leaking defects. Current industry best practice, however, is to utilize methods that do not involve direct welding onto the pipe. In particular, welded patches are not recommended by most pipeline codes owing to the risk of fatigue cracking at the fillet weld, and the increased stress concentration caused by the patch area. The process of weld deposition involves applying a series of weld passes over the damaged area, increasing the local wall thickness (see Fig. 7.7).

Snug-fit sleeve repair

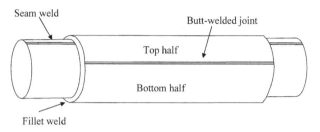

Figure 7.6   Snug-fit sleeve

7.4.9 Composite sleeve repair

The main type of composite sleeve repair is the 'clock spring' repair method, which is made from a high-strength fibreglass material. The procedure involves wrapping layers of material around the circumference of the pipe (see Fig. 7.8). In addition, bonding of the interface between the layers is made through a fast-curing adhesive material. This repair method is generally suited for blunt defects such as corrosion.

Offshore pipeline repairs are more complicated due to the divers' access requirements and operational considerations. Operational considerations include whether to flood the

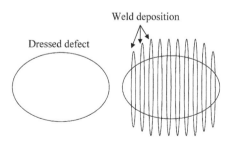

Figure 7.7   Weld deposition

Composite sleeve repair

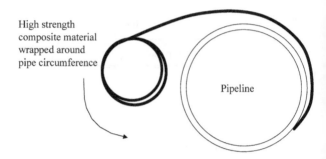

High strength
composite material
wrapped around
pipe circumference

Pipeline

Figure 7.8   Composite sleeve repair

pipeline with seawater or to isolate the damaged area using sealing pigs.

7.4.10 Mechanical clamps

If used as a permanent repair for an offshore pipeline, mechanical clamps are often filled with epoxy grout, which strengthens the pipeline. These devices are commonly used for repairing offshore dents caused by a dragged anchor – the epoxy mixture prevents any possible fatigue damage. A disadvantage is that these devices are usually bulky and heavy, particularly for large-diameter pipelines.

7.4.11 Offshore stopple and bypass operation

If a pipeline is severely damaged, such as a buckle caused by span deflection, the section of pipe may need to be replaced using a stopple and bypass operation. Similar to the onshore stopple and bypass operation, the damaged section of pipeline is removed and replaced. This is done using either:

- welding the new pipeline section in place (divers conduct the welding activities under water);
- use of mechanical connectors (here the new spool section

is connected to the existing pipeline using specialist connectors).

During a repair procedure, it is common for pipeline operators to reduce the operating pressure. Factors that influence the need for pressure reduction include:

- safety consideration (location specific);
- defect type and size;
- the current operating pressure.

Codes such as IGE/TD/1 recommend reducing the operating pressure to 30% SMYS [2].

Chapter 8

Pipeline Decommissioning and Industry Developments

8.1 Pipeline decommissioning

This book has covered the overall life cycle of a pipeline, looking at design, construction, operation and maintenance. The final consideration for a pipeline operator is decommissioning. Figure 8.1 shows a general curve representing the probability of failure during the life cycle of a pipeline. The probability of failure is high during the construction stage owing to the risk of damage by machinery or work personnel. The probability significantly decreases once the pipeline has been pressure tested and any threats to integrity have been identified. As the age of the pipeline increases and approaches its intended design life, the probability of failure increases again.

Finally, once a pipeline exceeds its design life, and is no longer cost effective to operate, it will eventually need to be decommissioned (i.e. removed from service). This process

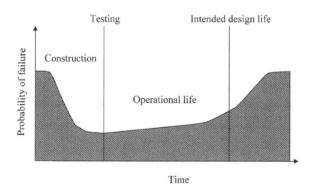

Figure 8.1   Failure probability versus time

involves safely removing the product from the pipeline and either:

- dismantling and removing the pipeline from the ground;
- sealing the ends of the pipeline and abandoning it.

In most cases the option chosen is completely to purge the pipeline of product and then seal it. For offshore pipelines this ensures that no remnant material leaks into the surrounding soil or water. For onshore pipelines, its route may affect roads, railways, river crossings and rural areas, hence appropriate landowners and authorities will need to be consulted. Abandoning a pipeline impacts upon the local environment, including:

- influence on the local water drainage;
- corrosion of the pipeline into the soil;
- subsidence, if the pipeline starts to deform owing to corrosion;
- risk of contamination if the product is not completely purged.

If a pipeline is to be abandoned, a risk assessment may be required to quantify the environmental and safety risks. Following decommissioning and abandonment, it is up to the pipeline operator to ensure that appropriate records are kept and that landowners are consulted. These records usually include:

- the previous route of the pipeline;
- the product type;
- information on the size, material type and thickness of the pipeline;
- abandonment procedures;
- any implemented safety precautions.

Guidance on decommissioning and abandonment is also provided in pipeline codes such as B31.4 [7], B31.8 [8] and IGE/TD/1 [2].

Abandonment of offshore pipelines requires the same key

decisions, in that the operator has to decide whether to leave the pipeline buried offshore or, if the pipeline is exposed, whether to remove the pipeline in sections so that it poses no hazard to ships or local marine life. There have also been instances where pipelines have been decommissioned and used to form artificial reefs, with the aim of assisting and developing the surrounding ecosystem.

8.2 Pipeline industry developments

The pipeline industry is continually changing, and developments occur in areas ranging from new legislation, construction methods, integrity and maintenance activities through to new repair techniques.

8.2.1 Pipeline codes and regulations

In response to the changing pipeline energy market in the United States, the Office of Pipeline Safety has issued new federal regulations (CFR) Title 49 Part 186-199 [**38**]. These new regulations cover safety programmes, transportation using gas pipelines, LNG facilities and transportation by pipelines containing other hazardous liquids. Part 192 applies to transportation of gas by pipelines and stipulates minimum federal safety standards. This places responsibility on the operator to implement integrity management programmes and to ensure that they:

- Utilize appropriate assessment and pipeline integrity activities;
- adopt regular inspection of the pipeline network;
- conduct ongoing assessment and maintenance.

Changes and updates have also occurred in some of the pipeline codes. These include the following:

- A supplement to the ASME B31.8 code – ASME B31.8S [**39**] – has been produced to give guidance on integrity and management practices.
- The British Standards Institute has published the current

version of BS 7910:2005 [**35**] which supersedes the previous version published in 1999.

8.2.2 Pipeline construction

Construction of pipelines has traditionally been from steel, but with new developments in material technology, new products have emerged in the market, such as composite material pipelines [**40**]. One new type of pipe that is available is the Bondstrand® GRE (glass-reinforced epoxy) system (see Fig. 8.2). This product was developed by the Ameron International Corporation and can be used for either oil or gas pipelines. Benefits of this new type of pipe include:

- The GRE pipe system weighs 12–25% of a comparable steel pipe.
- It performs well against internal corrosion.
- It resists external corrosion.
- The smooth internal surface reduces head loss.
- There are reduced installation and maintenance costs.
- It has an increased service life.

Unlike steel pipelines which require welding, GRE pipelines use a range of joining systems including adhesive, mechanical or threaded joints. Pipelines are often required to operate at high pressures, and this system is able to tolerate pressures of up to 245 bar. A further pipe system has been developed that allows pressures of up to 400 bar, known as the Bondstrand® SSL. This incorporates high-tensile steel in a glass-reinforced jacket. As pipeline operators aim to reduce costs associated with failures and maintenance activities, these systems provide a practical alternative to conventional steel pipelines.

For further information, contact Ameron BV, Fibreglass-Composite Pipe Group, PO Box 6, NL-4190 CA, Geldermalsen, The Netherlands. Email: info@ameron-fpg.nl.

Website: www.ameron-fpg.nl.

Bondstrand manufacturing process
Picture courtesy of Ameron B.V

**Figure 8.2   Bondstrand® GRE (glass-reinforced epoxy)
system**

8.2.3 Inspection technology

There are many different commercial companies that provide specialist inspection services. Inspection technology has changed over recent years, and currently most defects (corrosion, dents, stress corrosion cracking, etc.) can be accurately sized. One area of current research is on inspection of unpiggable pipelines, using techniques such as direct assessment. The following key elements are usually required for a successful pipeline inspection using pigs:

- low product flow velocity;
- bend radii above $3D$;
- constant diameter;
- access to pig traps.

In many instances, not all of the above requirements can be met owing to the pipeline configuration or requirement on the pipeline operator to maintain a high product flow. An example of one such problem is described below, and how Rosen Inspection Ltd has addressed a number of issues to develop a solution that allows inspection [41].

The project required a total of 1243 miles of pipeline to be inspected. Rosen Inspection were required by a pipeline operator to develop a tool that would negotiate $1.5D$ bends and significant elevation changes (4920 ft), and, because of pressure to maintain continuity of pipeline operation, the tool needed to tolerate a gas flowrate of 22.2 mile/h. In addition, the operator required a tool that would provide elevation profile, have active speed control, and use MFL technology.

The inspected pipelines were large at 48 and 56 in diameter. On account of these diameters and environmental conditions along the route, an improved defect location methodology was required. The designed solution was a compact single-body inspection tool equipped with an XYZ navigation system, active speed control and the ability to tolerate $1.5D$ bends. To improve manoeuvrability, the inspection tool was incorporated into a single body, with onboard units such as speed control, data storage equipment and electronics integrated as part of a miniaturized and compact design. Conventional inspection tools have a secondary array of sensors that discriminate between internal and external features, but in the present case this function was incorporated into a single array of sensors as an independent measurement channel (see Fig. 8.3). Speed control of the tool was achieved by designing a flow control valve that would keep the inspection tool speed within specified limits while the pipeline was operating at a high flowrate.

For further information, contact Rosen Technology, Am Seitenkanal 8, 49811 Lingen, Germany. Email: tbeuker@ roseninspection.net. Website: www.RosenInspection.net.

8.2.4 Repair techniques and pipeline live intervention

As described in Chapter 7, the current best practice in pipeline repair techniques is to utilize methods that do not involve direct welding onto the pipe. The problem with direct welding (T-connections, etc.) is that the internal pressure has to be reduced during this operation. An alternative approach has been developed by Advantica, known as the Grouted® tee connection. This new method does not require any site welding, allowing full flow to be maintained. It can also

Picture courtesy of Rosen inspection

Figure 8.3   MFL inspection tool

tolerate a larger amount of ovality. Figure 8.4 shows the layout of the grouted tee connection.

Some of the main benefits of this system are as follows:

- There is zero welding on the pipeline.

Picture courtesy of Advantica

Figure 8.4 Grouted™ tee connection

- It is simple to install.
- It maintains maximum flow during installation.
- It removes pipe ovality problems.
- It can be installed on thin-wall pipe with a large diameter/thickness ratio.
- Surface preparation only requires grit blasting.

For further information, contact Advantica, Ashby Road, Loughborough, Leicestershire LE11 3GR. Email: info.uk @advantica.biz. Website: www.advantica.biz.

References

1 PD 8010, Code of Practice for Pipelines: Part 2 – Pipelines on Land: Design, Construction and Installation, Published under the board of authority of BSI, London, 2004.

2 IGE/TD/1 Edition 4, Steel Pipelines for High Pressure Gas Transmission, The Institution of Gas Engineers, Communication 1670, London, 2001.

3 British Standard Business Information, *Storage Tanks, Piping and Pipelines, An International Survey of Design and Approval Requirements,* 2002 (BSI, London).

4 The Pipeline Industries Guild, *Pipeline: Design, Construction and Operation,* 1984, pp. 35 (Longman Inc., London).

5 **McAllister, E. W.** *Pipeline Rules of Thumb Handbook*, 6th edition, 2005, p. 314 (Gulf Professional Publishing, Elsevier, London).

6 PD 8010, Code of Practice for Pipelines: Part 3 – Pipelines Subsea: Design, Construction and Installation, Published under the board of authority of BSI, London, 2004.

7 ANSI/ASME B31.4, *Pipeline Transportation Systems for Liquid Hydrocarbons and Other Liquids,* 2002 (ASME, New York).

8 ANSI/ASME B31.8, *Gas Transmission and Distribution Piping Systems,* 2004 (ASME, New York).

9 API 5L, *Specification for Line Pipe, Exploration and Production Department,* 43rd edition, 2004 (American Petroleum Institute, Washington, DC).

10 CSA Z662, *Oil and Gas Pipeline Systems,* 4th edition, 2003 (Canadian Standards Associated, Ontario, Canada).

11 NACE Recommended Practice, RP0169 Control of External Corrosion on Underground or Submerged Metallic Piping Systems, NACE International, Houston, 2002.

12 *The Pipeline Safety Regulations, Health and Safety. Guidance on Regulations,* 1996 (HMSO, London).

13 **Eiber, R. J.** and **Jones, D.G.** An analysis of reportable incidents for natural gas transmission and gathering lines, June 1984 to 1990. AGA report NG-18, Report 200, Washington, DC, August 1992.

14 NACE Recommended Practice, RP0175 Control of Internal

Corrosion in Steel Pipeline and Piping Systems, Houston, 2002.

15 ANSI/ASME B31.G, *Manual for Determining Remaining Strength of Corroded Pipelines: Supplement to B31 Code-Pressure Piping*, 1991, pp. 1–13 (ASME, New York).

16 **Kiefner, J. F.** and **Vieth, P. H.** 'A modified criterion for evaluating the strength of corroded pipe, Final report for Project PR 3-805 to the Pipeline Supervisory Committee of the American Gas Association, Battelle, Ohio, 1989, pp. 15–18.

17 Det Norske Vareitas (DNV), Recommended Practice DNV-RP-F101, Corroded Pipelines, Hovik, Norway, October 2004.

18 **Stephens, R.** and **Francini, R. B.** A review and evaluation of the remaining strength criteria for corrosion defects in transmission pipelines. Battelle, Proceedings of ETCE/ OMAE2000 Joint Conference on *Energy for the New Millennium*, New Orleans, 14–17 February 2000, pp. 6–7.

19 **Stephens, D. R. , Leis, B. N. , Kurre, J. D.** and **Rudland, D. L.** Development of an alternative failure criterion for residual strength of corrosion defects in moderate to high toughness pipe, Battelle report to PRC International Report, AGA catalog number L51794, January 1999.

20 **Kastner, W.** Critical crack sizes in ductile piping. *Int. J. Pressure Vessels and Piping*, 1981, **9**, 197–219.

21 API 1156, *Effects of Smooth and Rock Dents on Liquid Petroleum Pipelines*, 1st edition, November 1997 (American Petroleum Institute, Washington, DC).

22 **Corder, I** and **Chatain, P.** EPRG recommendations for the assessment of the resistance of pipelines to external damage. EPRG/PRC 10th Biennial Joint Technical Conference on *Linepipe Research*, Cambridge, 18–21 April 1995, paper 12.

23 **Hopkins, P. Jones, D. G.** and **Clyne, A.** The significance of dents and defects in transmission pipelines. Conference C376/ 049, Institution of Mechanical Engineers, London, November 1989, pp. 137–140,

24 **Eiber, R. J., Maxey, W. A., Bert, C. W.,** and **McCure, G. M.** The effects of dents on the failure characteristics of linepipe, Battelle Columbus Laboratories, NG-18, Report 125, AGA catalogue number L51403, May 1981.

25 **Wang, K. C.** and **Smith, E. D.** Exploratory pressure testing of 168 mm diameter pipes with dents, Canadian Centre for

Mineral and Energy Technology (CANMET), Canada, Report ERP/PMRL 79–81 (TR), December 1979.

26 **Wang, K. C.** and **Smith, E. D.** The Eefect of mechanical damage on fracture initiation in linepipe Part I – Dents, Canadian Centre for Mineral and Energy Technology (CANMET), Canada, Report ERP/PMRL 82–11 (TR), January 1982.

27 **Jones, D. G.** The significance of mechanical damage in pipelines. *3R Int.*, July 1982, **21**(7).

28 **Hopkins, P., Jones, D. G.** and **Clyne, A. C.** The significance of dents and defects in transmission pipelines. In Proceedings of International Conference on *Pipework, Engineering and Operation*, Institution of Mechanical Engineers, London, February 1989, paper C376/049, pp. 137–140.

29 **Corder, I.** and **Chatain, P.** EPRG recommendations for the assessment of the resistance of pipelines to external damage. In Proceedings of the EPRG/PRC 10th Biennial Joint Technical Meeting on *Line Pipe Research*, Cambridge, UK, April 1995, pp. 12–14.

30 **Hopkins, P.** The significance of mechanical damage in gas transmission pipelines. In Proceedings of EPRG/NG-18 8th Biennial Joint Technical Meeting on *Line Pipe Research*, Paris, France, 14–17 May 1991, Vol. 2, paper 25.

31 **Alexander, C. R.** and **Kiefner, J. F.** Effects of smooth and rock dents on liquid petroleum pipelines, Final report to The American Petroleum Institute, Stress Engineering Services, Inc., and Kiefner and Associates, Inc., 10 October 1997, API Publication 1156, 1st Edition, Washington, DC, November 1997.

32 **Fowler, J. R., Alexander, C. R., Kovach, P. J.** and **Connelly, L. M.** Cyclic pressure fatigue life of pipelines with plain dents, dents with gouges, and dents with welds, Final report to the Pipeline Research Committee of the American Gas Association, Report PR-201-927 and PR-201-9324, Stress Engineering Services, Inc., Washington, DC, June 1994.

33 PD 5500 British Standard, *Specification for Unfired Fusion Welded Pressure Vessels*, 2000 (BSI, London).

34 Deutsche Norm, Design of Steel Pressure Pipes, DIN 2413 Part 1, Iron and Steel Institute, London, October 1993.

35 BS 7910, *Guide on Methods for Assessing the Acceptability of Flaws in Metallic Structures*, 2005 (BSI, London).

36 API Recommended Practice 579, *Fitness-for Service*, 1st edition, January 2000 (American Petroleum Institute, Washington, DC).

37 API 1160, *Managing System Integrity for Hazardous Liquid Pipelines*, 1st Edition, August 2001 (American Petroleum Institute, Washington, DC).

38 Code of Federal Regulations CFR Title 49: Transportation, Parts 186–199, Office of Pipeline Safety, December 2005.

39 ASME B31.8S, *Managing System Integrity of Gas Pipelines*, 2004 (American Society of Mechanical Engineers, New York).

40 GRE piping: an alternative to metal and thermoplastic products. *Tube and Pipe Technol. – Plastic and Composite*, November/December 2005, 72–73.

41 Re-assessment survey of large-diameter pipelines using compact multi-purpose in-line inspection tools'. *J. Pipeline Integrity*, **4**(3), 155–162.

Index

Index

Index